Heterogeneous Catalysis in Organic Transformations

Emerging Materials and Technologies

Series Editor: *Boris I. Kharissov*

For more information about this series, please visit: https://www.routledge.com/Emerging-Materials-and-Technologies/book-series/CRCEMT

Heterogeneous Catalysis in Organic Transformations

Edited by

Varun Rawat
Anirban Das
Chandra Mohan Srivastava

CRC Press
Taylor & Francis Group
Boca Raton London New York

CRC Press is an imprint of the
Taylor & Francis Group, an **informa** business

Cover image: Andrii Muzyka/Shutterstock

First edition published 2022
by CRC Press
6000 Broken Sound Parkway NW, Suite 300, Boca Raton, FL 33487-2742

and by CRC Press
2 Park Square, Milton Park, Abingdon, Oxon, OX14 4RN

CRC Press is an imprint of Taylor & Francis Group, LLC

© 2022 Taylor & Francis Group, LLC

Library of Congress Cataloging-in-Publication Data
Names: Rawat, Varun, editor. | Das, Anirban (Professor of chemistry), editor. | Srivastava, Chandra Mohan, editor.
Title: Heterogeneous catalysis in organic transformations / edited by Varun Rawat, Anirban Das, and Chandra Mohan Srivastava.
Description: First edition. | Boca Raton : CRC Press, 2022. | Series: Emerging materials and technologies | Includes bibliographical references and index. | Summary: "This book provides a complete description of role of heterogeneous catalysis in organic transformations and offers a review of the current and near future technologies and applications. Providing a comprehensive examination of heterogeneous catalysis from the basics through recent advances, this book will be of keen interest to undergraduates, graduates, and researchers in chemistry, chemical engineering, and associated fields"-- Provided by publisher.
Identifiers: LCCN 2021049203 (print) | LCCN 2021049204 (ebook) | ISBN 9780367647872 (hardback) | ISBN 9780367647889 (paperback) | ISBN 9781003126270 (ebook)
Subjects: LCSH: Heterogeneous catalysis. | Organic compounds--Synthesis.
Classification: LCC QD505 .H466 2022 (print) | LCC QD505 (ebook) | DDC 541/.395--dc23/eng/20211216
LC record available at https://lccn.loc.gov/2021049203
LC ebook record available at https://lccn.loc.gov/2021049204

ISBN: 978-0-367-64787-2 (hbk)
ISBN: 978-0-367-64788-9 (pbk)
ISBN: 978-1-003-12627-0 (ebk)

DOI: 10.1201/9781003126270

Typeset in Times
by SPi Technologies India Pvt Ltd (Straive)

Contents

Contents

Preface

Heterogeneous catalysis is the backbone of the chemical industry. Pharmaceutical, petrochemicals, fine chemicals, polymer, and many other industries rely on heterogeneous catalytic processes. Even a slight increase in the selectivity and yield of these transformations can save billions of dollars for some industries. Thus, a detailed understanding of the reaction mechanisms is sought to design catalysts or tune reaction conditions for better yield and selectivity for the desired products. Broadly, the study of heterogeneous catalysts involves:

1. Design, synthesis, and appropriate characterization of catalysts,
2. Carrying out the catalytic transformation and record of yield and selectivity, and
3. Investigation of mechanistic aspects of these transformations.

Based on inputs obtained from mechanistic investigations, the design and synthesis of catalysts or catalyst systems are fine-tuned to obtain higher yields and desired selectivity.

The advent of nanocatalysts has introduced tremendous versatility into the field. The tremendous increase in surface area of these nanosized catalysts as compared to bulk catalysts of the same composition usually results in a huge (sometimes several orders of magnitude) increase in reaction rates. Additionally, nanomaterials may have different surface terminations than their bulk counterparts, which can have a profound effect on the catalytic activity. The surface area and surface termination may also be tailored by variation in these catalysts' shape, size, and morphology. With the evolution of synthetic methodology, the shape, size, and morphology of these catalysts may now be functionalized with relative ease. Different parts (e.g., edges, corners, shapes) of the same nanoparticle may result in different products. Quantum confinement effects add an additional dimensionality to this field. This book introduces the reader to the variety of catalysts used in organic transformations with specific emphasis on points 1–3 described above.

The first chapter gives a general introduction to heterogeneous catalysts in organic transformations. The second chapter describes the reported organic transformations on oxide nanoparticles. These particles are one of the most extensively used catalysts, as most of them are inexpensive to produce. Noble metals were considered quite unreactive till about 50 years ago, but the development of nanoparticles of these metals opened up new avenues of organic transformations. Organic transformations on Au, Pb, Ir, Pt, and Ag nanoparticles or their composites are described in the third chapter. The fourth chapter describes organometallic compounds that catalyze organic transformations, while the fifth chapter describes solid-supported catalysts. These supported catalysts are very significant as many homogeneous catalysts act as heterogeneous catalysts when immobilized on supports. Apart from imparting stability, synergistic effects between the support and the immobilized catalysts impart unique properties to these catalysts. Mesoporous materials described in chapter 6 are another class of materials that are being intensively investigated as heterogeneous catalysts. These catalysts' pore sizes and shapes can be modified to selectively encapsulate desired substrates and thus impart the desired selectivity. Chapter 7 describes the role of metal-based heterogeneous catalysts.

We hope that the book will cater to the needs of graduate students in chemistry and help them get a flavor of the rich and immensely interesting field of organic transformations using heterogeneous catalysts.

Varun Rawat
Haryana, India

Anirban Das
New Delhi, India

Chandra Mohan Srivastava
Haryana, India

Editors

Dr Varun Rawat earned his MSc Chemistry degree from the University of Delhi and PhD from CSIR-National Chemical Laboratory, India. He then joined the group of Professor Arkadi Vigalok as a PBC postdoctoral fellow, where he worked on the synthesis of calixarene-based complexes. He is currently an Assistant Professor of Chemistry at Amity University Haryana, India, and a visiting scientist at Tel Aviv University, Israel. His research interest includes the synthesis and application of calixarene-based chemosensors and catalysts. He has published over 20 peer-reviewed papers, five patents, and five book chapters.

Dr Anirban Das earned his BSc Chemistry (Hons) degree from the University of Delhi, India. He then earned an MS degree in Inorganic Chemistry from the University of Toledo, Ohio, USA (Prof. Mark Mason, 2007), and a PhD in Inorganic Chemistry at University of Idaho, USA (Prof. Chien M Wai, 2012). He thereafter held postdoctoral positions in the United States at Northwestern University (Prof. Eric Weitz, 2012–2014), University of California (Prof. Francisco Zaera, 2014–2015) and at the Wayne State University (Prof. Eranda Nikolla, 2015–2017). Thereafter, he joined as a CSIR Pool Scientist at IIT Delhi, India working with Prof. A. K. Ganguli (2017–2020). He assumed his current position at the Department of Chemistry, Biochemistry and Forensic Sciences at Amity University Haryana as an Associate Professor in 2020. His current interests are broadly in the study of reaction mechanisms and chemical and photoelectrocatalysis using nanomaterials.

Dr Chandra Mohan Srivastava earned his M. Tech. in Plastics Engineering and PhD degree from Central Institute of Plastics Engineering & Technology and Delhi Technological University, Delhi respectively. He has worked as a Lecturer at Central Institute of Plastics Engineering & Technology, Ahmedabad. Currently he is an Assistant Professor of Chemistry in Amity School of Applied Sciences, Amity University Haryana, India. His research interests include polymer, biopolymer, nano-materials for biomedical applications, nanofabrication using electrospinning, and polymer composites. He has published 25 peer-reviewed papers, six patents, and five book chapters.

Contributors

Anirban Das
Amity School of Applied Sciences
Amity University Haryana
Gurugram (Haryana), India

Anamika Srivastava
Department of Chemistry
Banasthali Vidyapith
Jaipur (Vanasthali), India

Dr. Ashutosh Sharan Singh
Assistant Professor
Department of Chemistry
Maharishi Markandeswar Deemed to be
    University

Deepika
Department of Chemistry
Chaudhary Devi Lal University
Sirsa (Haryana), India

Dipti Vaya
Amity School of Applied Sciences
Amity University Haryana
Gurugram (Haryana), India

Garima Sachdeva
Amity School of Applied Sciences
Amity University Haryana
Gurugram (Haryana), India

Gyandshwar Kumar Rao
Amity School of Applied Sciences
Amity University Haryana
Gurugram (Haryana), India

Jyoti Dhariwal
Amity School of Applied Sciences
Amity University Haryana
Gurugram (Haryana), India

Jyotirmoy Maity
Department of Chemistry
St. Stephen's College,
University of Delhi
Delhi, India

Kamal Nayan Sharma
Amity School of Applied Sciences
Amity University Haryana
Gurugram (Haryana), India

Komal
Amity School of Applied Sciences
Amity University Haryana
Gurugram (Haryana), India

Laxmi Devi
Amity School of Applied Sciences
Amity University Haryana
Gurugram (Haryana), India

Manish Srivastava
Department of Chemistry
Banasthali Vidyapith
Jaipur (Vanasthali), India

Meenal Batra
Department of Chemistry
Banasthali Vidyapith
Newai (Rajasthan), India

Monika Vats
Amity School of Applied Sciences
Amity University Haryana
Gurugram (Haryana), India

Monu Verma
Department of Environmental
    Engineering
The University of Seoul
Seoul, Republic of Korea

Navjeet Kaur
Department of Chemistry
Banasthali Vidyapith
Newai (Rajasthan), India

Pooja Rawat
Department of Applied Physics and
    Institute of Natural Sciences
Kyung Hee University
Yong-In, Gyong-gi, Republic of
    Korea

Sudesh Kumar
Department of Chemistry
Banasthali Vidyapith
Jaipur (Vanasthali), India

Sukhdev Singh
iSm2 *Stereo*, Aix-Marseille University
Marseille, France

Sunita Kanwar
Amity School of Applied Sciences
Amity University Haryana
Gurugram (Haryana), India

Vanshika Singh
Department of Chemistry
Banasthali Vidyapith
Newai (Rajasthan), India

Varun Rawat
Amity School of Applied Sciences
Amity University Haryana
Gurugram (Haryana), India

Ved Prakash Verma
Department of Chemistry
Banasthali Vidyapith
Newai (Rajasthan), India

1 Introduction to Heterogeneous Catalysis in Organic Transformation

Garima Sachdeva, Gyandshwar Kumar Rao, and Varun Rawat
Amity University Haryana, Gurugram (Haryana), India

Ved Prakash Verma and Navjeet Kaur
Banasthali Vidyapith, Newai (Rajasthan), India

CONTENTS

DOI: 10.1201/9781003126270-1

1.1 INTRODUCTION

The concept of the term "Catalysis" was first described by Berzelius in 1835 [1] and later defined scientifically by Ostwald in 1894 [2]. However, catalysts have been used for thousands of years and the oldest example that we still encounter today is the fermentation process which the Egyptians first discovered to produce wine [3]. Catalysis has been the backbone of industrial applications and is used extensively in manufacturing agro- and petrochemicals, cosmetics, pharmaceuticals and medicines, polymers, aliments, and many more [4]. By definition, a catalyst is a substance that can increase a chemical reaction's speed without undergoing any change itself. Figure 1.1 gives a very simplistic view of the energetics involved in a catalyzed reaction. A catalyst can change the reaction pathways that lower its activation energy and increase the rate of reaction. However, a catalyst does not alter the equilibrium of a reaction, which means that the product formation is achieved at a faster rate, whereas the yield of the reaction remains unaffected. A catalyst undergoes a reversible chemical change, and it regenerates its previous form at the end of a chemical cycle.

   In many organic transformations without catalysts, either the reactions do not occur or they proceed very slowly. For example, the addition of water to an alkene proceeds to completion if an acid is added in a catalytic amount, leading to the

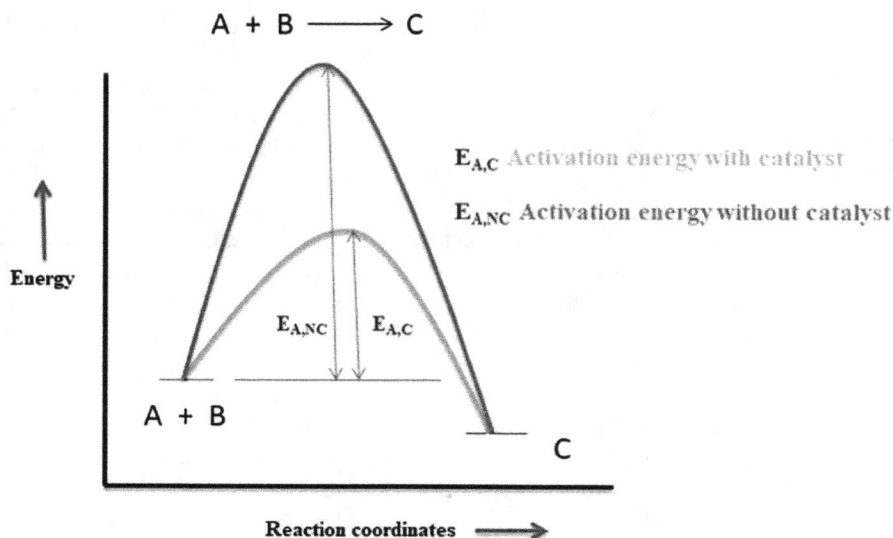

FIGURE 1.1   Energy profile diagram of a reaction with and without catalyst.

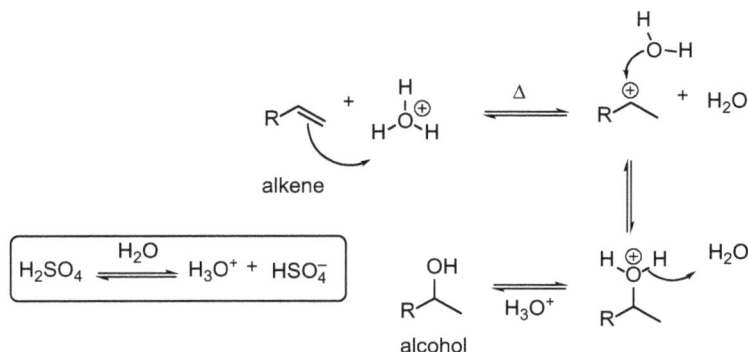

FIGURE 1.2   Hydration of alkenes in the presence of an acid.

formation of an alcohol. Without the catalyst, the reaction does not proceed at all. This reaction involves breaking a pi-bond in the alkene and an O–H bond in water and the formation of C–H and C–OH bonds. By the addition of an acid, the concentration of hydronium ions (H_3O^+) will increase in the solution, which acts as a catalyst for the reaction. The H_3O^+ ions then react with the alkene molecule to form an alkyl cation according to Markovnikov's rule (Figure 1.2). The alkyl cation has a strong tendency for water molecules, and even though water molecules are weak nucleophiles, they rapidly attack the alkyl cation forming alcohol and leaving protons to regenerate H_3O^+ molecules. Without an acid catalyst, the neutral water molecule would not be electrophilic enough for the pi-bond to attack it, and thus, the reaction would not proceed [5].

1.2   TYPES OF CATALYSTS

In general, on the basis of their phase in a reaction, catalysts are divided into two broad categories – homogeneous and heterogeneous. However, the other classifications also include heterogenized homogeneous catalysts and biocatalysts (Figure 1.3). Their brief accounts are summarized as follows:

1.2.1   HOMOGENEOUS CATALYST

A homogeneous catalyst has the same phase in the reaction mixture as that of reactants. The high homogeneity of the same phase of reactants and catalyst results in their high interactions, leading to high reactivity and selectivity under mild reaction conditions. Some important instances of homogeneous catalysts include Brønsted and Lewis acids, transition metal complexes, organometallic complexes, and organocatalysts. Some notable examples of chemical reactions that use homogeneous catalysts are carbonylation, oxidation, hydrocyanation, metathesis, hydrogenation, and C–H and C–C activation/functionalization [6].

```
                                    ┌──────────────┐
                                    │  CATALYSIS   │
                                    └──────────────┘
```

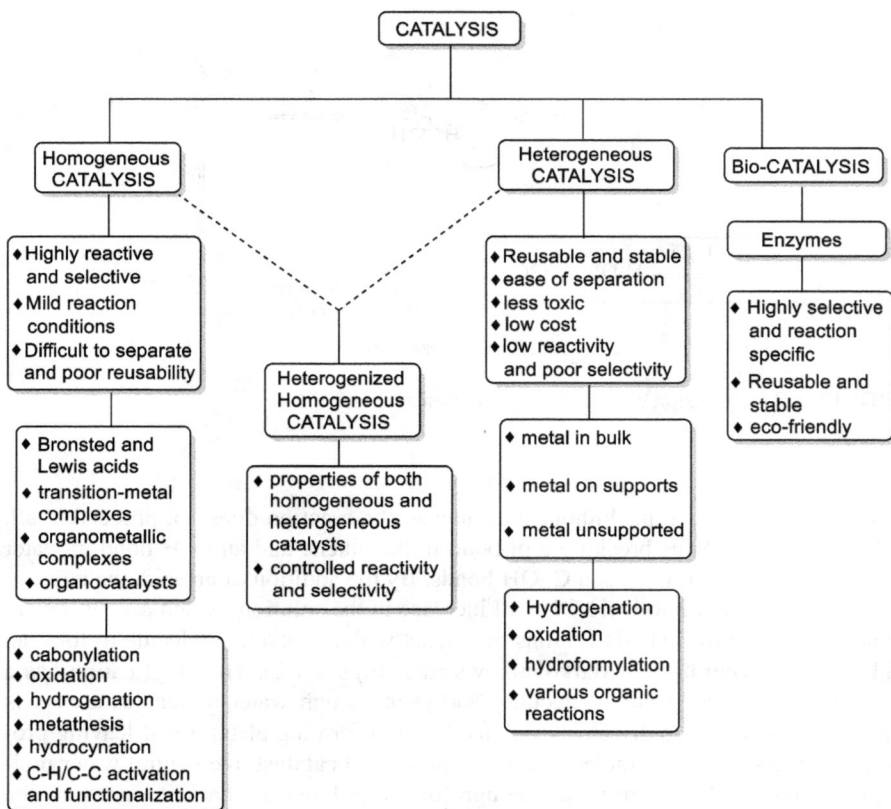

FIGURE 1.3 Classification, properties, and usage of catalysts.

1.2.2 HETEROGENEOUS CATALYST

As the name suggests, the catalysts exist in a different phase than reactants in the reaction mixture [7]. For example, heterogeneous catalysts are used in the Haber–Bosch process for the ammonia synthesis (iron as a catalyst) and Fischer–Tropsch process to produce hydrocarbons (transition metal catalysts used such as iron, cobalt, ruthenium, and nickel). The recovery, reusability, and easy separation of products from the reaction mixture make heterogeneous catalysts the first choice for industrial use. These catalysts (primarily metals) can be used as fine particles, powders, granules, deposited on the solid surface (supported catalysts), or used in bulk form.

1.2.3 HETEROGENIZED HOMOGENEOUS CATALYSTS

The complex nature of heterogeneous catalysts prevents their analysis and characterization at a molecular level, making their development difficult through structure–reactivity relationships. Additionally, the traditional heterogeneous catalysts are less reactive and show poor selectivity in a chemical reaction. In order to overcome these issues, homogeneous catalysts are often embedded on solid surfaces to add heterogeneity in their nature. This approach brings features of both homogeneous (selectivity and reactivity) and heterogeneous (reproducibility) catalysts together in one catalyst, which greatly enhance the outcome of a reaction. Heterogeneity can be obtained by immobilizing catalysts on the solid surface *via* surface processes like physisorption or chemisorption [8].

1.2.4　BIOCATALYSTS

The reactions in our biological systems are carried out by natural biocatalysts, which are primarily enzymes. Enzymes are natural proteins and can be used to catalyze very specific chemical reactions in a laboratory. They are isolated from animal or plant tissues and microbes such as yeast, bacteria, or fungi. Biocatalysts are notable for their great selectivity and efficiency, as well as for their environmental friendliness and gentle reaction conditions. Nowadays, biocatalysts are an alternative to conventional industrial catalysts. Immobilizing these enzymes on solid supports turns enzymes into heterogeneous solid catalysts, enhancing their activity and stability. It also increases their lifetime, and they can be recycled for many usages [9]. The last two categories of catalysts classification are used arbitrarily and can be merged into other types depending upon the usage and nature of the catalyst.

1.3　ORIGIN OF HETEROGENEOUS CATALYSIS

The evolution of heterogeneous catalysts took place by hit and trial methods. For instance, the progress in ammonia synthesis in the early 1900s demonstrated how empirical screening could yield an excellent catalytic process [10]. Although the worldwide ammonia production was enhanced with this discovery, its mechanism was unknown for an extended period until Gerhard Ertt discovered and improved the process by introducing promoters [11]. Moreover, his work also concentrated on many other heterogeneous catalytic procedures like catalytic oxidation of carbon monoxide (CO) over platinum. For his seminal work on examining chemical processes on a solid surface, he was awarded the Nobel Prize in chemistry in 2007. The complexity of a catalyst can range from well-defined supported metal nanoparticles to millimeter-sized multicomponent catalyst bodies with various often distinct activities. Heterogeneous catalysts are still extensively utilized in the production of bulk chemicals, but now they are also used for the selective synthesis of intermediates and fine chemicals. This has been made feasible because of significant advances in catalyst designing at the molecular level, like combining the benefits of nanostructured solids and grafted organometallic complexes.

1.4　COMPARISON BETWEEN HOMOGENEOUS AND HETEROGENEOUS CATALYSIS

It is well known that depending on the chemical phase of the catalyst and reactants, different forms of catalysis can be distinguished. In homogeneous catalysis, the catalysts work in the same phase as the reactants and are often found in the liquid phase. It has been seen that mechanistic studies of homogeneous catalysts are comparatively

TABLE 1.1

Advantages and Disadvantages of Different Catalyst Systems

Catalyst	Advantages	Disadvantages
1. Homogeneous Catalyst	➤ The catalyst gets mixed with the reaction mixture, which permits high degree of interaction between the catalyst and reactant. ➤ Homogeneous catalysts are more active in many reactions. ➤ Time required is less and high yield is observed. ➤ Usually more selective for single product.	➤ Homogeneous catalysts have less thermal stability. ➤ Cannot be separated from the reaction mixture and hence can't be reused.
2. Heterogeneous Catalyst	➤ Catalyst is easy to separate after the reaction completion. ➤ Possess a long catalytic life. ➤ Can be regenerated and reused.	➤ Any type of coating on the surface can reduce the activity. ➤ Only effective on the surfaces.

easier than heterogeneous catalysts. The examples are Fe-porphyrin complexes active for the oxidation, Zn-complexes for decarboxylation reaction, Cu-imidazole (from histidine) complexes in hemocyanin, etc. [12].

In heterogeneous catalysis, the catalyst is in a different phase than reactants. In most of the reactions, the catalyst is solid, and reactants are in gas or liquid phases; hence, the rate-determining step occurs at the solid surface of the catalyst. For example, the development of catalytic converters for automobiles benefitted from acknowledging the reaction kinetics for CO oxidation and NOx on Pt single catalysts [13]. Table 1.1 gives a brief outline of the advantages and disadvantages of the two main catalyst systems.

1.5   CONTRIBUTION OF HETEROGENEOUS CATALYSIS

Heterogeneous catalysts are often the backbone of chemical and energy industries, making it one of the most rapidly developing branches of chemistry. Although the basic reaction mechanism is frequently well understood, substantial research is usually required to improve the efficiency of a reaction or the selectivity of a catalyst. Based on the research outcomes of a reaction profile and knowing the role of a catalyst, it becomes possible to decide whether a process can be moved from a laboratory scale to a plant scale. Designing powerful and active catalytic processes is the goal of all laboratories and industries in the world. For the production of valuable chemicals, catalytic processes play an essential role, and most energy-dependent methods rely on heterogeneous catalysis. Consequently, close to 80% of all chemical reactions are carried out using some form of heterogeneous reaction [13].

In heterogeneous catalysis, a phase boundary separates the reactants and catalysts. Most of the essential and large-scale industrial processes such as the contact process for producing sulfuric acid, Haber–Bosch process for ammonia production, Ostwald's process for making nitric acid, and many more depend on heterogeneous catalysts.

It has been observed that noble metals like Pt, Pd, Rh, Ni, Co, and Ir are preferred as heterogeneous catalysts [14]. They have been widely employed in the petrochemical sector, medication manufacture, and environmental protection, but they are unable to meet the rising demand due to high costs and scarcity of such noble metals in nature. The main aim of studying heterogeneous catalysis is to gain knowledge of the mechanism at the molecular level and design the catalyst with a preferred active site. In heterogeneous catalysis, at least one reacting species must be chemisorbed to react catalytically on the solid surface; hence, chemisorption is the elementary step for the atoms or molecules to react [15]. The prime advantage of heterogeneous catalysts is that they can perform their function in a broad pH spectrum [16]. In heterogeneous catalysis, the chemical reaction occurs at the material's surface; hence, heterogeneous catalysts are highly porous materials with a large surface area.

Heterogeneous catalysis is the primary approach used to convert petroleum and natural gas into environmentally benign fuels and produce alternatives like hydrogen fuel and biofuels. For example, biodiesel is produced with the help of heterogeneous metal oxides because of their selectivity, ease of separation of catalyst from the reaction mixture, and fewer processing steps [17]. Calcium oxide (CaO) is the most extensively used solid basic catalyst among the several metal oxides known because it has many advantages, including long catalytic life, high activity, and just moderate reaction conditions. It should be specified that the recovery process of heterogeneous catalysts is cheap and can be done smoothly.

1.6 MECHANISM OF HETEROGENEOUS CATALYSIS

Irving Langmuir proposed a theory for adsorption on solid surfaces in the early 1900s, which is one of the foundations of contemporary surface chemistry. Langmuir's approach is based on the hypothesis that only a single layer of molecules (monolayer) is formed, making the surface inactive against further adsorption when completely covered. Furthermore, the adsorbed particles should not interact with one another. Langmuir's hypothesis introduced the concept of adsorbed particles consuming adsorption sites, laying the groundwork for the present atomistic understanding of surface chemistry [18]. In accordance with the adsorption theory of catalysis, reactants in the gaseous state or dissolved state in the solution get adsorbed on the surface of a solid catalyst.

At least one of the reactants adsorbs on catalyst surfaces containing coordination-unsaturated atoms during heterogeneous catalytic reactions and is activated; adsorbed reactants then conduct surface reactions to generate adsorbed products that subsequently desorb products from the catalyst surfaces. Consequently, there will be an increase in the concentration of reactants on the catalyst surface that enhances the chances of reaction between two species, and hence the rate of the reaction increases. When the reactants get attached to the catalyst surface, some amount of the energy is released, which in turn increases the rate of reaction. A brief schematic representation of heterogeneous catalysis is shown in Figure 1.4. Each step possesses a different rate, and the slowest one is the rate-determining step. The catalytic reaction rate limit depends on the number of reactants in contact with the surface but is independent of the surface kinetic properties.

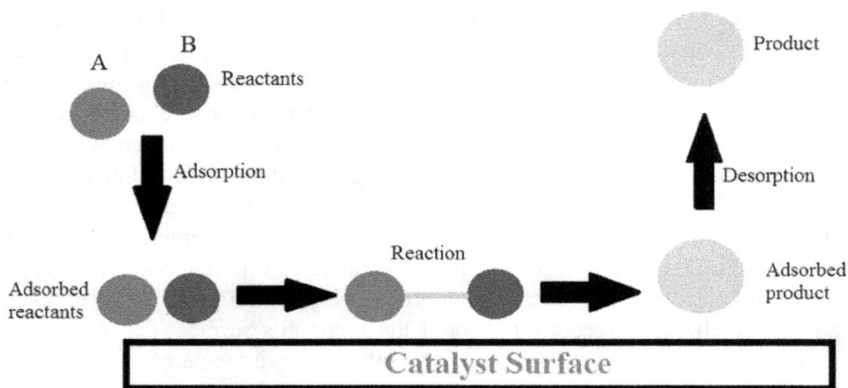

FIGURE 1.4   A schematic representation of the different steps in heterogeneous catalysis.

1.6.1   LANGMUIR–HINSHELWOOD MECHANISM

The first and most important type of mechanism explaining the surface chemical reactions is the Langmuir–Hinshelwood mechanism, which Langmuir first proposed in 1921 and further refined by Hinshelwood in 1926 [19]. The Langmuir–Hinshelwood mechanism has been acknowledged as the favored mechanism for the great majority of surface catalytic reactions. The Langmuir–Hinshelwood mechanism is comprised of the following steps: adsorption from the gas phases, molecule dissociation at the surface, adsorption–adsorption interactions, and desorption to the gas phase. This mechanism includes a bimolecular reaction involving two adsorbed molecules on the surface. After the reaction, it is also expected that the product would depart the surface (desorb) (Figure 1.5(I)).

1.6.2   ELEY–RIDEAL MECHANISM

A second important reaction mechanism was proposed by Eley and Rideal [20]. The Eley–Rideal mechanism explains a reaction between a chemisorbed reactant and a non-chemisorbed reactant. An Eley–Rideal reaction is distinguished by the fact that one of the reactants is not chemisorbed locally and so not in equilibrium with the surface. According to this mechanism, only one of the reactants is adsorbed on the surface in this process, and it combines immediately with an incoming molecule from the gas phase, bypassing the need for an adsorption site. Since, the gas-phase temperature does not have to be the same as that of the surface, the reaction can be classified as non-thermal (Figure 1.5(II).

1.6.3   HARRIS–KASEMO MECHANISM

In Harris–Kasemo type reactions, it is hypothesized that one of the reactants is chemisorbed while the other is trapped in a heated precursor state with a long lifetime. The Harris–Kasemo reaction has been discovered to be the dominant mechanism for the reaction of gas-phase atomic hydrogen with adsorbed hydrogen on Pt(III) [21].

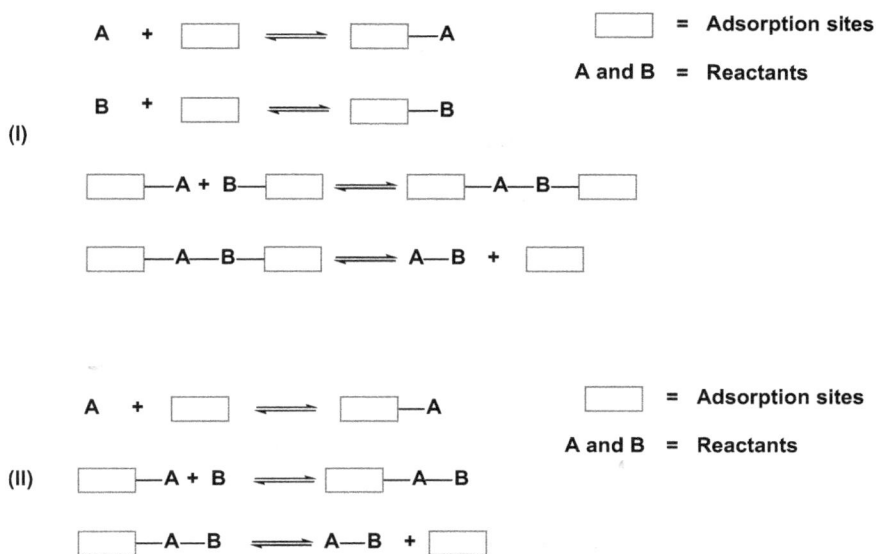

FIGURE 1.5 (I) A representative reaction scheme for the Langmuir–Hinshelwood mechanism. (II) A representative reaction scheme for the Eley–Rideal mechanism.

1.7 CATEGORIZATION OF HETEROGENEOUS CATALYSTS

Heterogeneous catalysts are usually divided into two categories based on the active site [22].

1.7.1 CATALYSTS HAVING BASIC SITES

Catalysts having basic sites include CaO, MgO, ZnO, and waste materials infused with NaOH or KOH. This kind of catalyst can catalyze transesterification reaction under mild conditions, and generally high yield of biodiesel is obtained from vegetable oil or animal fat. However, due to the basic properties of the catalyst, it is prone to the free fatty acids (FFA) content in oil, resulting in soap formation instead of biodiesel. Another limitation is the leaching of active catalyst sites, which causes product contamination and deactivation of the catalyst.

1.7.2 CATALYSTS HAVING ACIDIC SITES

Catalysts having acidic sites include SO_4^{2-}/ZrO_2, SO_4^{2-}/TiO_2, SO_4^{2-}/SnO_2, zeolites, sulfonic ion exchange resins, and sulfonated carbon-based catalysts. These catalysts are less prone to FFA content in oil and bring about esterification and transesterification reactions simultaneously. Nevertheless, the reaction rate is relatively slow, so extreme conditions are required, such as high temperature (more than 100 °C) with alcohol to oil molar ratio more than 12:1 to speed up the rate of the reaction.

1.8   CHARACTERIZATION TECHNIQUES NEEDED
FOR HETEROGENEOUS CATALYSTS

Characterization is an essential step required for catalyst development. Structure elucidation, composition, and chemical properties of solids used in heterogeneous catalysis, adsorbate, and intermediates present on the surface of catalysts during the reaction are important to understand the relation between properties of catalysts and their catalytic functioning. Here we will discuss the most common spectroscopy techniques and methods used to characterize heterogeneous catalysts [23].

1.8.1   X-RAY DIFFRACTION

X-Ray Diffraction (XRD) is usually used to determine the bulk structure and composition of heterogeneous catalysts with a crystalline structure. Diffraction can provide rather direct information on the periodicity or translational symmetry of a surface structure. While this information is useful, it does not provide a clear image of where the surface atoms are located inside the unit mesh. To extract this information, we must examine the diffracted beam intensities and connect them to a quantitative scattering theory, just as we do in X-ray crystallography of bulk structures. It is commonly confined to the identification of specific lattice planes which produce peaks at their corresponding angular position 2θ, which is resolved by Bragg's Law ($2d \sin \theta = n\lambda$) [24].

1.8.2   X-RAY ABSORPTION SPECTROSCOPY

X-ray absorption spectroscopy (XAS) is frequently utilized for determining the local geometric and electrical structure of materials. Typically, the experiment is conducted using synchrotron radiation, which generates powerful and tunable X-ray beams. Samples can be in gas, solution, or solid form. This technique can be utilized for the structural and compositional analysis of solid catalysts. The main advantage of using this technique is that it does not require long-range order in the samples and can work efficiently under non-vacuum conditions [25].

1.8.3   ELECTRON MICROSCOPY

An electron microscope is a type of microscope that uses a beam of accelerated electrons as a source of illumination. Since an electron's wavelength can be much shorter than that of visible light photons, as a consequence, electron microscopes have a higher resolving power and can depict the structure of small particles. This method is used to create high-resolution pictures of biological and non-biological specimens. Electron microscopy is a recognized standard tool for nanomaterial characterization and can be used for determining the shape and size of solid catalysts [26]. It can be done in two modes: SEM (scanning electron microscopy) and TEM (transmission electron microscopy). The transmission electron microscope (TEM), the first-ever type of electron microscope, uses a large electron beam to visualize specimens and creates an image of the material. The scanning electron microscope (SEM) generates images by probing the sample with a concentrated electron beam that scans through a rectangular section of the sample. SEM is effective for visualizing catalyst particles (composition) and surfaces (topography) with micrometer diameters, but TEM is helpful for obtaining detailed structural information.

1.8.4 NUCLEAR MAGNETIC RESONANCE

Nuclear magnetic resonance spectroscopy (NMR) and magnetic resonance imaging (MRI) are excellent tools for the in-depth evaluation of reaction parameters. NMR is the most common technique used to analyze reactions in the liquid state, but in magic-angle spinning mode, and it can also be used to characterize the solid catalyst, particularly zeolites. Because of its ability to provide atomic-level perspicacity into the structure, interaction, and dynamics of molecules by confirming connectivity and proximity between identical or distinct nuclei, solid-state NMR correlation spectroscopy is becoming increasingly recognized as a powerful tool in the study of catalysts and catalytic reactions. For example, the ^{29}Si NMR signal is used for the determination of the coordination environment of Si in the zeolite structure [27].

1.8.5 ELECTRON SPIN RESONANCE MICROSCOPY

Electron Spin Resonance (ESR) also called paramagnetic resonance (EPR) is employed in heterogeneous catalysis to study the paramagnetic behavior of species having one or more unpaired electrons. The necessary use of ESR in catalysis is for the coordination chemistry of transition metal cations consolidated into zeolites or metal oxides [28]. ESR or EPR has a wide range of applications in the physical, chemical, biological, and geological sciences. Although ESR microscopy's applicability in materials research is presently limited, the technique's potential capability in the study of catalysis is immense.

1.9 SIGNIFICANCE OF HETEROGENEOUS CATALYSIS

Heterogeneous catalysis is the support system for many chemical and energy industries and is developing very rapidly. They provide an advantage to social and economic sectors and help obtain sustainable energy to benefit the environment. Heterogeneous catalysts increase the reaction rate with less cost and reduce the usage of chemicals in industries, hence working in an eco-friendly and greener way [29]. Every branch of chemistry is directly or indirectly dependent on heterogeneous catalysts. They can be operated under vigorous conditions, which are very much important to produce a wide range of chemicals on which the modern community depends.

The purpose of this book is to provide the reader with a comprehensive overview of the scientific and practical aspects of heterogeneous catalysts. We hope to give a brief yet broad outline of various heterogeneous catalysts and their applications in many organic reactions. Several different types of heterogeneous catalysts have been used for different purposes, which are the subject of several book chapters of this book. Application areas of heterogeneous catalysis have had a significant impact on academic research and industrial research.

REFERENCES

1. Berzelius, A. A. (1835). Sur un Force Jusqu'ici Peu Remarquée qui est. Probablement Active Dans la Formation des Composés Organiques, Section on Vegetable Chemistry, *Jahres-Bericht*, *14*, 237.
2. Ostwald, W. (1894). Definition der katalyse, *Zeitschrift für Physikalische Chemie*, *15*, 705–706.
3. Beker, C. (2018). From Langmuir to Ertl: The "Nobel" History of the Surface Science Approach to Heterogeneous Catalysis. *Encyclopedia of Interfacial Chemistry*, (pp. 99–106), Elsevier, Amsterdam.
4. (a) Armor, J. N. (2011). A history of industrial catalysis. *Catalysis Today*, *163*, 3–9. (b) Ma, Z. & Zaera, F. (2005). *Encyclopedia of Inorganic Chemistry©*, John Wiley & Sons, Ltd, New York. (c) Ma, Z., & Zaera, F. (2006). Heterogeneous Catalysis by Metals. *Encyclopedia of Inorganic Chemistry*. https://onlinelibrary.wiley.com/doi/abs/10.1002/9781119951438.eibc0079.pub2.
5. Resch, V., & Hanefeld, U. (2015). The selective addition of water. *Catalysis Science & Technology*, *5*(*3*), 1385–1399.
6. (a) Parshall, G. W. (1978). Industrial applications of homogeneous catalysis: A review. *Journal of Molecular Catalysis*, *4*(*4*), 243–270. (b) Cornils, B., & Herrmann, W. A. (2003). Concepts in homogeneous catalysis: The Industrial view. *Journal of Catalysis*, *216*(*1–2*), 23–31. (c) Duca, G. (2012). *Homogeneous Catalysis with Metal Complexes*, (pp. 1–484), Springer Series in Chemical Physics, Springer Nature, Switzerland. (d) Van Leeuwen, P. W. N. M. (2004). *Homogeneous Catalysis*, (pp. 1–412), Kluwer Academic Publishers, Netherlands.
7. Wacławek, S., Padil, V. V. T., & Černík, M. (2018). Major advances and challenges in heterogeneous catalysis for environmental applications: A review. *Ecological Chemistry and Engineering S*, *25*(*1*), 9–34.
8. (a) Choplin, A., & Quignard, F. (1998). From supported homogeneous catalysts to heterogeneous molecular catalysts. *Coordination Chemistry Reviews*, *178*, 1679–1702. (b) Collis, A. E. C., & Horváth, I. T. (2011). Heterogenization of homogeneous catalytic systems. *Catalysis Science & Technology*, *1*(*6*), 912–919.
9. (a) Hughes, G., & Lewis, J. C. (2018). Introduction: Biocatalysis in industry. *Chemical Reviews*, *118*(*1*), 1–3. (b) López-Gallego, F., Jackson, E., & Betancor, L. (2017). Heterogeneous systems biocatalysis: The path to the fabrication of self-sufficient artificial metabolic cells. *Chemistry – A European Journal*, *23*(*71*), 17841–17849.
10. (a) Boudart, M. (1994) Ammonia synthesis: The Bellwether reaction in heterogeneous catalysis. *Topics in Catalysis*, *1*(*3–4*), 405–414. (b) Schlögl, R. (2003). Catalytic synthesis of Ammonia – A "Never-EndingStory"? *Angewandte Chemie International Edition*, *42* (*18*), 2004–2008.
11. Ertl, G. (1983). Surface characterization of ammonia synthesis catalysts. *Journal of Catalysis*, *79*(*2*), 359–377.
12 Reedijik, J., & Bowwman, E. (1999). *Bioinorganic Catalysis*, M. Dekker, Inc.: New York.
13. (a) Védrine, J. C. (2018) *Védrine Metal Oxides in Heterogeneous Oxidation Catalysis: State of the Art and Challenges for a More Sustainable World*, Wiley-VCH Verlag GmbH & CoKGaA. Weinheim. Védrine, J. C. (2019). Metal oxides in heterogeneous oxidation catalysis: state of the art and challenges for a more sustainable world. *ChemSusChem*, *12*, 577–588. doi:10.1002/cssc.201802248. (b) Taylor, K. C. (1984). Automobile Catalytic Converters. *Catalysis*, (pp. 119–170), Springer, Berlin, Heidelberg.
14. Boudart, M. (1985). Heterogeneous catalysis by metals. *Journal of Molecular Catalysis*, *30*(*1–2*), 27–38.

15. Bond, G. C. (1974). *Heterogeneous Catalysis: Principles and Applications*, Clarendon Press, Oxford.
16. Wacławek, S., Lutze, H. V., Grübel, K., Padil, V. V. T., Černík, M., & Dionysiou, D. D. (2017). Chemistry of persulfates in water and wastewater treatment: A review. *Chemical Engineering Journal*, *330*, 44–62.
17. (a) Borges, M. E., & Díaz, L. (2012). Recent developments on heterogeneous catalysts for biodiesel production by oil esterification and transesterification reactions: A review. *Renewable and Sustainable Energy Reviews*, *16(5)*, 2839–2849. (b) Math, M. C., Kumar, S. P., & Chetty, S. V. (2010). Technologies for biodiesel production from used cooking oil—A review. *Energy for Sustainable Development*, *14(4)*, 339–345.
18. Langmuir, I. (1916). The evaporation, condensation and reflection of molecules and the mechanism of adsorption. *Physical Review*, *8(2)*, 149–176.
19. (a) Langmuir, I. (1922). Part II.—"Heterogeneous reactions". Chemical reactions on surfaces. *Transactions of the Faraday Society*, *17*, 607–620. (b) Hinshelwood, C. N. (1926). On the theory of unimolecular reactions. *Proceedings of the Royal Society*, *113*, 230–233.
20. (a) Rideal, E. K.. (1939). A note on a simple molecular mechanism for heterogeneous catalytic reactions. *Proceedings of the Cambridge Philosophical*, *35*, 130–132. (b) Eley, D. D. (1948). The catalytic activation of hydrogen. *Advances in Catalysis*, *1*, 157–199.
21. Harris, J., & Kasemo, B. (1981). On precursor mechanisms for surface reactions. *Surface Science Letters*, *105(2–3)*, L281–L287.
22. Lam, M. K., & Lee, K. T.. Scale up and commercialization of algal cultivation and Biofuel production. *Biofuels from Algae*, 261–286).
23. (a) Thomas, J. M., & Thomas, W. J. (1997). *Principles and Practice of Heterogeneous Catalysis*, VCH, Weinheim. Muhler, M., Thomas, J. M., & Thomas, W. J. (1997). Principles and Practice of Heterogeneous Catalysis, VCH, Weinheim. D. M. *Berichte Der Bunsengesellschaft Für Physikalische Chemie*, *101(10)*, 1560–1560. (b) Santen, R. A. V., Leeuwen, P. W. N. M., & Moulijn, J. A. (1999). In *Catalysis: An Integrated Approach*. B. A. Averill (Eds.), Elsevier, Amsterdam. (c) Anderson, R. B. (1976). In *Experimental Methods in Catalytic Research*, Vols. II & III. P. T. Dawson (Eds.), Academic Press, New York. (d) Delgass, W. N., Haller, G. L., Kellerman, R., & Lunsford, J. H. (1979). *Spectroscopy in Heterogeneous Catalysis*, Academic Press, New York. (e) Thomas, J. M. (1980). In *Characterization of Catalysts*. R. M. Lambert (Eds.), Wiley, Chichester. (f) Dekker, M. (1984). In *Characterization of Heterogeneous Catalysts*. F. Delannay (Ed.), Marcel Dekker Inc., New York. (g) *Spectroscopic Characterization of Heterogeneous Catalysts* (1990). J. L. G. Fierro (Ed.), Elsevier, Amsterdam. (h) Imelik, B.. (1994). *Catalyst Characterization: Physical Techniques for Solid Materials*. J. C. Vedrine (Eds.), Plenum Press, New York. (i) Niemantsverdriet, J. W. (2000). *Spectroscopy in Catalysis: An Introduction*, (2nd ed.), Wiley-VCH, Weinheim. (j) *n-Situ Spectroscopy in Heterogeneous Catalysis* (2002). J. F. Haw (Ed.), Wiley-VCH, Weinheim. (k) *In-Situ Spectroscopy of Catalysts* (2004). B. M. Weckhuysen (Ed.), American Scientific Publishers, Stevenson Ranch. (l) *Methods of Surface Analysis: Techniques and Applications* (1992). J. M. Walls (Ed.), Cambridge University Press, Cambridge. (m) Woodruff, D. P., & Delchar, T. A. (1994). *Modern Techniques of Surface Science*, (2nd ed.), Cambridge University Press, Cambridge.
24. (a) Jellinek, M. H., & Fankuchen, I. (1948). *Advances in Catalysis*, *1*, 257. (b) Klug, H. P., & Alexander, L. E. (1954). *X-Ray Diffraction Procedures for Polycrystalline and Amorphous Materials*, Wiley, New York. (c) Paulus, E. F., & Gieren, A. (2001). In *Handbook of Analytical Techniques*, H. Gunzler, & A. Williams (Eds.), (p. 373), Wiley-VCH, Weinheim.

25. (a) Nordstrand, R. A. (1960). *Advances in Catalysis*, *12*, 149. (b) Sinfelt, J. H., Via, G. H., & Lytle, F. W. (1984). *Catalysis Reviews Science and Engineering*, *26*, 81. (c) *X-Ray Absorption Fine Structure for Catalysts and Surfaces* (1996). Y. Iwasawa (Ed.), World Scientific, Singapore.
26. (a) Amelinckx, S., Dyck, D., & Landuyt, J. V. (1997). In *Handbook of Microscopy: Applications in Materials Science, Solid-State Physics and Chemistry*. G. V. Tendeloo (Eds.), VCH, Weinheim. (b) Amelinckx, S., Dyck, D. V., Landuyt, J. V. & Tendeloo, G. V. (1999). *Electron Microscopy: Principles and Fundamentals*, VCH, Weinheim.
27. (a) Fyfe, C. A., Feng, Y., Grondey, H., Kokotailo, G. T., & Gies, H. (1991). *Chemical Reviews*, *91*, 1525. (b) Klinowski, J. (1993). *Analytica Chimica Acta*, *283*, 929. (c) Engelhardt, G. (1997). In *Handbook of Heterogeneous Catalysts*, Vol. 2, G. Ertl, H. Knözinger, & J. Weitkamp (Eds.), (p. 525), VCH, Weinheim. (d) Bell, A. T. (1999). *Colloids and Surfaces A*, 221. (e) Grey, C. P. (2003). In *Handbook of Zeolite Science and Technology*, S. M. Auerbach, K. A. Carrado, & P. K. Dutta (Eds.), (p. 205), Marcel Dekker, New York.
28. (a) Dyrek, K., & Che, M. (1997). EPR as a tool to investigate the transition metal chemistry on oxide surfaces. *Chemical Reviews*, *97(1)*, 305–332. (b) Labanowska, M. (2001). EPR monitoring of redox processes in transition metal oxide catalysts. *ChemPhysChem*, *2(12)*, 712.
29. Ali, M. E., Rahman, M. M., Sarkar, S. M., & Hamid, S. B. A. (2014). Heterogeneous metal catalysts for oxidation reactions. *Journal of Nanomaterials*, 2014, 1–23.

2 Oxide Nanoparticles in Heterogeneous Catalysis

Garima Sachdeva, Jyoti Dhariwal, Monika Vats, and Varun Rawat
Amity University Haryana, Gurugram (Haryana), India

Manish Srivastava and Anamika Srivastava
Banasthali Vidyapith, Jaipur (Vanasthali), India

CONTENTS

DOI: 10.1201/9781003126270-2

2.1   INTRODUCTION

Nanomaterials are structures, devices, or particles of size ranging from 1 to 100 nanometers and are made up of carbon, metal, metal oxides, or inorganic matter. Nanomaterials can be classified based on the number of dimensions of a material, which are outside the nanoscale (<100 nm) range; based on this parameter, they may be classified as zero-dimensional (0D), one-dimensional (1D), two-dimensional (2D), or three dimensional (3D). Accordingly, in zero-dimensional (0D) nanomaterials, all the dimensions are measured within the nanoscale. One dimension lies outside the nanoscale in one-dimensional nanomaterials (1D). Nanotubes, nanorods, and nanowires all fall within this category. Two dimensions are outside the nanoscale in two-dimensional nanomaterials (2D). Graphene, nanofilms, nanolayers, and nanocoatings are examples of this class, which have plate-like structures. Materials that are not confined to the nanoscale in any dimension are referred to as three-dimensional nanomaterials (3D). This class contains bulk powders, dispersions of nanoparticles, bundles of nanowires, and nanotubes, as well as multi-nanolayers [1]. Nanoparticles (NPs) are three-layered molecules with a complicated structure that includes a surface layer, which can be functionalized with various small molecules, surfactants, polymers, and metal ions. The shell layer, in every way, is a chemically distinct substance and the core, which is the most central part of NPs. In the last few decades, there has been massive progress in the field of nanotechnology. The results show the exceptional performance of nanoparticles (NPs) as heterogeneous catalysts in terms of selectivity, reactivity, and improved yields [2]. In addition, the high surface-to-volume ratio of NPs provides a larger number of active sites per unit area in comparison with their heterogeneous bulk counterparts [3]. NPs can also be broadly divided into various categories depending on their composition (Table 2.1) [4].

Various methods can be employed for the synthesis of NPs, but these methods are broadly divided into two main classes: bottom-up approach (A) and top-down approach (B). These approaches are further divided into various subclasses based on the operation, reaction conditions, and adopted protocols [5].

(A) Bottom-up approach includes

1. Spinning synthesis
2. Template support synthesis
3. Plasma or flame spraying synthesis
4. Laser pyrolysis
5. Chemical vapor deposition (CVD)

TABLE 2.1

Categorization of Nanoparticles

S. No.	Category	Sub-classes	Properties
1	Organic NPs	Polymeric-based NPs	These nanoparticles are biodegradable, non-toxic, and are sensitive to thermal and electromagnetic radiation such as heat and light. They are most widely used in the biomedical field, for example, drug delivery system. Dendrimers, micelles, liposomes, and ferritin are common examples of organic nanoparticles.
2	Inorganic NPs	Metal-based	These nanoparticles are synthesized from metals either by destructive or constructive methods. The commonly used metals for nanoparticle synthesis are aluminum (Al), cadmium (Cd), cobalt (Co), copper (Cu), gold (Au), iron (Fe), lead (Pb), silver (Ag), and zinc (Zn). Due to localized surface Plasmon resonance (LSPR) characteristics, these NPs possess unique optoelectrical properties.
		Metal oxide-based	Metal oxide nanoparticles are synthesized mainly to modify the properties of their respective metal-based nanoparticles. The commonly synthesized NPs are aluminum oxide (Al_2O_3), cerium oxide (CeO_2), iron oxide (Fe_2O_3), magnetite (Fe_3O_4), silicon dioxide (SiO_2), titanium oxide (TiO_2), and zinc oxide (ZnO).
3	Carbon-based NPs	Fullerenes	Fullerenes (C_{60}) is a carbon molecule that is spherical in shape and made up of pentagonal and hexagonal arranged carbon units, in which carbon atom is present in sp^2 hybridization. They possess high electrical conductivity, high strength, and electron affinity.
		Graphene	Graphene is an allotrope of carbon. Graphene is a hexagonal network of honeycomb lattice made up of carbon atoms in a two-dimensional planar surface. Generally, the thickness of the graphene sheet is around 1 nm.
		Carbon Nanotubes (CNT)	CNTs are elongated, tubular structure, 1–2 nm in diameter. They are widely synthesized by a chemical vapor deposition (CVD) technique. Due to their unique physical, chemical, and mechanical characteristics, these materials are used in nanocomposites for many commercial applications such as fillers, efficient gas adsorbents for environmental remediation, and as support medium for different inorganic and organic catalysts.
			It is an amorphous material made up of carbon, generally spherical in shape with diameters 20–70 nm.

6. Atomic or molecular condensation
7. Biological synthesis *via* bacteria, yeast, fungi, algae, plants, etc.
8. Sol–gel method
9. Citrate nitrate method
10. Hydrothermal synthesis

(B) Top-down approach includes

1. Mechanical milling
2. Chemical etching
3. Sputtering
4. Laser ablation
5. Electro-explosion

2.2 APPLICATION OF NPS

Considering the unique properties and ease of synthesis of NPs as discussed in the previous section, NPs can be used in a variety of applications. Some important applications are listed below.

2.2.1 APPLICATIONS IN DRUGS AND MEDICATIONS

Nanoparticles can deliver medications in the optimal dosage range; hence, they have received increasing interest from every discipline of medicine. Their use often results in higher therapeutic efficiency, decreased adverse effects, and improved patient compliance [6].

2.2.2 APPLICATIONS IN MANUFACTURING AND MATERIALS

Nanocrystalline materials are fascinating materials for materials science because their properties differ in size from their bulk counterparts. NPs have physicochemical features that produce unique electrical, mechanical, optical, and imaging qualities that are highly sought after in medicinal, commercial, and environmental applications [7].

2.2.3 APPLICATIONS IN THE ENVIRONMENT

The interaction of environmental pollutants with NPs is dependent on the NP characteristics, like size, composition, morphology, porosity, and aggregate structure. Superparamagnetic iron oxide NPs are effective sorbent materials for removing toxic metals like lead, thallium, cadmium, and arsenic from natural water [8].

2.2.4 APPLICATIONS IN ELECTRONICS

One-dimensional semiconductors and metals have unique structural, optical, and electrical capabilities, making them the main structural block for a new generation of electronic, sensor, and photonic materials. The ability to incorporate NPs in electrical, or optical devices, such as "bottom-up" or "self-assembly" techniques, is a crucial feature of NPs [9].

2.2.5 APPLICATION IN CATALYSIS

Organic processes catalyzed by metal/metal oxide nanoparticles (NPs) have recently received much interest. Nanoparticle catalysts are an effective and promising family of innovative heterogeneous catalysts because of their particular benefits, such as high catalytic activity, superior recyclability, and simple product purifications. Nanoparticles can be used as a catalyst in both homogeneous and heterogeneous processes. Recent examples of nanocatalysts include copper, ruthenium, rhodium, silver, palladium, iron, gold, nickel, and platinum nanoparticles and supports on silica, clays, zeolite, alumina, or carbon materials. Few selective nanoparticle-catalyzed reactions are listed below, which highlights the importance of nanoparticles in organic synthesis.

  i.   Coupling reactions
 ii.   Electrochemical reactions
iii.   C–H activation
 iv.   Photocatalytic reactions
  v.   Fine chemical synthesis
 vi.   Heterocyclic synthesis

2.3 COPPER OXIDE NANOPARTICLE-CATALYZED ORGANIC REACTIONS

Copper is the cheapest and abundant coinage metal found in the Earth's crust. Copper-based particles are found to enhance the reactions because of the variable oxidation states. CuO NPs can be synthesized *via* aqueous precipitation methods, where copper acetate is the precursor and NaOH is used as a stabilizing agent. The unique and remarkable properties of copper-based nanoparticles make them suitable for various organic transformations discussed below.

2.3.1 CARBON–CARBON BOND FORMATION

Carbon–carbon bond formation is one of the elemental transformations of synthetic organic chemistry. There are many reactions through which C–C bond can be formed, some of which are described below:

2.3.1.1 Stille Coupling Reaction

The Stille cross-coupling reaction is one of the most essential carbon–carbon bond formation reactions in organic synthesis due to the significance of the products formed, which play a crucial role as an intermediate in synthetic organic reactions and as structural elements in many biologically active compounds [10, 11].

Zhang et al. reported the use of Cu_2O nanoparticles in Stille coupling reaction between aryl halides and organotin reagent in the presence of tetrabutylammonium bromide (TBAB) and potassium fluoride as a promoter. In addition, the Cu_2O NPs/P(o-tol)$_3$/TBAB reagent system can be reused and restored at least three times without losing the activity in the coupling reaction. It has been observed that the system was quite adequate for the less active aryl chloride as well [12].

2.3.1.2 Arylation of Active Methylene Compounds

Kidwai and the group described the C-arylation of an active methylene compound. Treatment of iodobenzene with acetylacetone in the presence of DMSO solvent and CuO nanoparticle as a heterogeneous catalyst afforded 3-phenylpentane-2,4-dione in 80% yield (Figure 2.1). Other copper salts like $Cu(OAc)_2$ and bulk CuO were also utilized, but the product yield was very poor. The efficiency of CuO nanoparticles toward the arylation of active methylene compounds is due to the high surface area. Moreover, the catalyst can be reused for at least four cycles without losing the activity [13].

2.3.2 CARBON–NITROGEN BOND FORMATION

The carbon–nitrogen bond is one of the ample bonds found in organic chemistry in which a covalent bond exists between carbon and nitrogen. Nitrogen-containing heterocyclic compounds are one of the most sought-after compounds owing to their varying biological activity.

2.3.2.1 N-Arylation Reaction

Punniyamurthy et al. reported the use of recyclable CuO NPs to catalyze the C–N cross-coupling reaction of an amine with an aryl halide. The reaction of aniline and iodobenzene with CuO NPs in the presence of DMSO and KOH under air afforded the desired product in a 95% yield. Catalyst is inexpensive, reusable, and stable toward air [14]. Different solvents (DMSO, toluene, dioxane, and DMF) and base (KOH, K_2CO_3) were employed, but it was observed that the best yield could be realized with a combination of DMSO as solvent and KOH as the base. It was further reported that anilines with electron-donating groups (EDGs) were more effective those with electron-withdrawing groups (EWGs) in terms of reactivity.

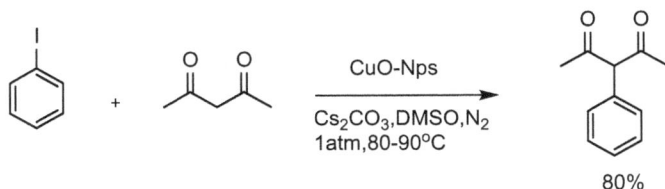

FIGURE 2.1   C-Arylation of active methylene compounds.

2.3.2.2 Arylation of Aromatic Heterocycles

Li et al. had introduced an effective methodology for N-arylation of nitrogen-containing heterocycles, particularly imidazole, indoles, and triazoles with aryl and heteroaryl halide using a cubic Cu_2O NPs/1,10-phenanthroline catalytic system and TBAF (tetra-n-butylammonium fluoride) as a base under solvent-free conditions (Figure 2.2). The group has performed the N-arylation with four Cu_2O systems, namely as a bulky compound, octahedral, spherical, and cubic nanoparticulate forms, but the excellent yield was obtained with Cubic Cu_2O nanoparticles. Hence, it has been observed that both particle size and shape of Cu_2O affected the yield of the corresponding product [15].

2.3.2.3 Amidation of Aryl Iodides

Punniyamurthy and co-workers reported the use of Cu_2O nanoparticles in PEG (polyethylene glycol) for the amidation of aryl iodide. To optimize the reaction condition a number of reactions were carried out with different molecular weights of PEG; one of the most important roles of PEG was to enhance the formation and stabilization of NPs. N-arylation of benzamide with 1-iodo-4 methylbenzene afforded the desired product in 53%, and the reaction is devoid of any external chelating ligand (Figure 2.3) [16]. Aryl halides involving both EWG and EDG were able to afford the desired product in high yield. It was also revealed that no leaching of Cu_2O occurred; hence, it can be recycled and reused easily without loss of activity.

2.3.2.4 Intramolecular Carbon–Nitrogen Bond Formation

Benzimidazoles are an important class of organic compounds because of their use in therapeutic and biological activities. Punniyamurthy et al. developed the methodology for synthesizing substituted benzimidazoles via intramolecular cyclization in the presence of CuO nanoparticles under air and ligand-free conditions in excellent yield (Figure 2.4). From an industrial point of view, this methodology is effective and

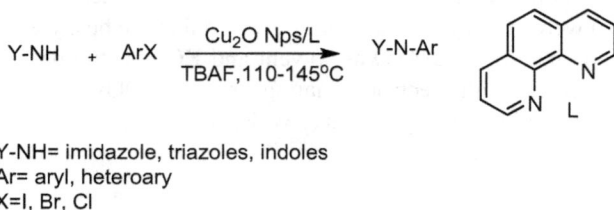

Y-NH + ArX $\xrightarrow[\text{TBAF,110-145°C}]{\text{Cu}_2\text{O Nps/L}}$ Y-N-Ar

Y-NH= imidazole, triazoles, indoles
Ar= aryl, heteroary
X=I, Br, Cl

FIGURE 2.2   Arylation of heterocycles.

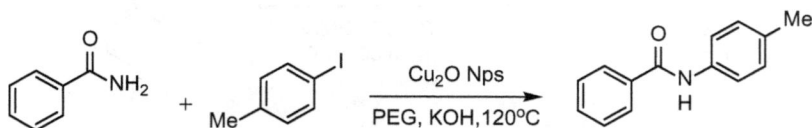

FIGURE 2.3   N-Arylation of benzamide with aryl iodide.

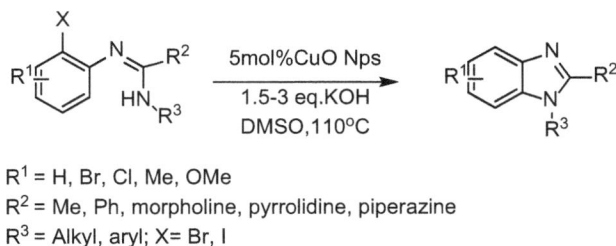

R^1 = H, Br, Cl, Me, OMe
R^2 = Me, Ph, morpholine, pyrrolidine, piperazine
R^3 = Alkyl, aryl; X= Br, I

FIGURE 2.4 Synthesis of substituted benzimidazoles.

eco-friendly as E-factor was decreased [17, 18]. The catalyst can be regained without loss in its selectivity and reactivity.

2.3.3 CARBON–OXYGEN BOND FORMATION

Polar covalent bond exists between carbon and oxygen. The C–O bond formation occurs via many mechanisms.

2.3.3.1 Oxygen Arylation

Punniyamurthy et al. described the simple and effective methodology for the C–O cross-coupling reaction between hydroxy compounds and iodobenzene by using CuO nanoparticles in the presence of KOH base and DMSO solvent at moderate temperature conditions. Corresponding products were obtained in high yield under ligand-free conditions. It has been seen that the substrate having EDG exhibited good reactivity in comparison with EWG [19]. This group also studied the reactions with phenyl tosylate, bromobenzene, phenylboronic acid, and chlorobenzene, but all gave the cross-coupled products in lower yields in comparison to iodobenzene.

2.3.4 CARBON–SULFUR BOND FORMATION

Carbon–sulfur bond is quite similar to carbon–oxygen bond; it is also a polar covalent bond due to its electronegativity difference. The carbon–sulfur (C–S) bond formation holds a prominent position in the race for the synthesis of valuable chemical entities, because organosulfur compounds are widely present in nature and various biological systems.

2.3.4.1 Carbon–Sulfur Cross-Coupling Reaction between Thiols and Iodobenzene

Technique for making C–S bonds is essential in synthetic organic chemistry. Conventional ways require harsh conditions for C–S bond formation; keeping this in mind, Punniyamurthy et al. were the first to utilize the CuO nanoparticles for the C–S cross-coupling reaction of aryl and alkyl thiols with iodobenzene. The reaction was carried out in DMSO solvent and KOH base under nitrogen atmosphere to obtain the desired product [20]. Different solvent systems like isopropyl alcohol, DMF, toluene,

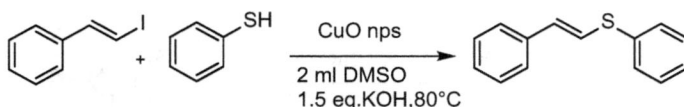

FIGURE 2.5 Synthesis of vinyl sulfides.

1,4-dioxane were also tested for the C–S cross-coupling reactions, which were found to be ineffective.

Like other coupling reactions, this reaction also proceeds through an oxidative addition followed by reductive elimination. After completion of the reaction, the catalyst was recovered by using the centrifugation technique.

2.3.4.2 Stereoselective Synthesis of Vinyl Sulfides Catalyzed by using CuO Nanoparticles

Vinyl sulfides are seen in many biologically active molecules and possess a wide range of applications in organic reactions. Rao and his group were the first to report the cross-coupling reaction of vinyl halides with thiols catalyzed by CuO nanoparticles under ligand-free conditions in the presence of KOH (Figure 2.5). The corresponding stereoselective vinyl sulfides were produced in excellent yield by using a recyclable catalyst [21]. The group also investigated the effect of solvent and base, and it was found that DMF gave the moderate yield, but water and toluene were ineffective. The methodology is free from using moisture-sensitive systems and external ligands.

2.3.5 CARBON–SELENIUM BOND FORMATION

Diaryl selenides act as an important reagent in catalysis and organic synthesis. Rao et al. reported the use of CuO nanoparticles for the formation of carbon–selenium bonds. Treatment of aryl halide with diaryl diselenide under ligand-free conditions afforded diaryl selenide in high yield. Catalyst is environmentally benign, inexpensive, reusable, and effective [22].

2.4 ZINC OXIDE NANOPARTICLE-CATALYZED ORGANIC REACTIONS

Zinc oxide (ZnO) nanoparticles work as an effective catalyst for various types of organic reactions due to their eco-friendly nature, ease to handle property, recyclability, and cost-effectiveness. They show high catalytic activity due to the large surface area. Some of the organic reactions catalyzed by ZnO nanoparticles are discussed in this section. ZnO NPs are easy to prepare and show activity in a multitude of reactions.

2.4.1 SYNTHESIS OF NITROGEN-CONTAINING HETEROCYCLES

The analogs of nitrogen-based heterocycles occupy an exclusive position as a valuable source of therapeutic agents in medicinal chemistry. More than 75% of drugs approved by the FDA and currently available in the market are nitrogen-containing heterocyclic moieties. Due to this reason many research groups have focused on the synthesis of these important organic compounds using heterogeneous catalysis.

2.4.1.1 Synthesis of Polysubstituted Pyrroles

Pyrroles belong to an important class of heterocyclic compounds and are extensively used as bioactive molecules in alkaloids, co-enzymes, and porphyrins. Sabbaghan et al. reported the effective and eco-friendly method for synthesis of polysubstituted pyrroles via three-component reaction of amine, dialkyl acetylene dicarboxylates, and phenacyl bromide using ZnO-nanorod as catalyst under solvent-free conditions (Figure 2.6). Solvent-free conditions make the synthesis simple, efficient in terms of atom economy, and prevent solvent toxicity and hazards.

Group has also utilized nanosheets as a catalyst, but the yield obtained was not significant; hence it was concluded that the catalytic activity depends on the size and morphology of ZnO [23]. Low-temperature conditions in the synthesis of catalyst and pyrrole preparation are fascinating from an economic point of view.

2.4.1.2 Synthesis of Benzimidazoles

Benzimidazoles are one of the essential natural and heterocyclic compounds. Alinezhad et al. had described a proficient method for synthesizing benzimidazole derivatives in the presence of ZnO NPs as a catalyst from formic acid and o-phenylenediamines under solvent-free conditions (Figure 2.7) [24]. The ability of different solvents like acetonitrile, water, and dichloromethane was also checked for the reaction, and they afforded lower product yield and more by-products compared to solvent-free conditions. The protocol did not use expensive reagents and mild reaction conditions, which makes it practical and eco-friendly.

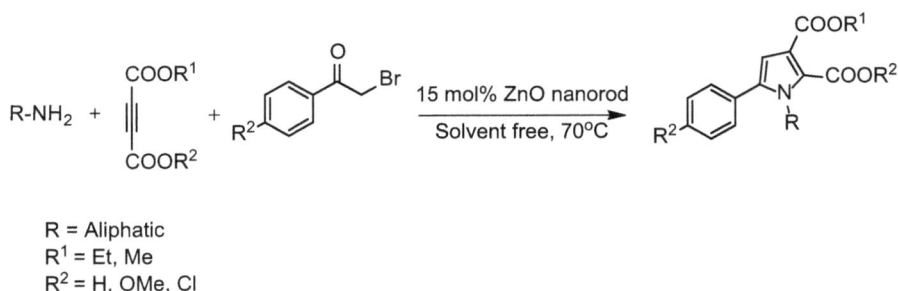

R = Aliphatic
R^1 = Et, Me
R^2 = H, OMe, Cl

FIGURE 2.6 Synthesis of polysubstituted pyrroles.

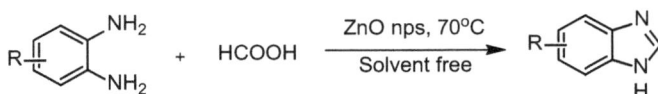

FIGURE 2.7 Synthesis of benzimidazoles.

2.4.1.3   Synthesis of Imidazo-fused Polyheterocycles

Shrivastava et al. had developed an eco-friendly and straightforward method for the synthesis of biologically compatible pyrazole-coupled imidazo[1,2-α]pyridine derivatives via one-pot three-component reaction of alkyl-4-formyl-1-phenyl-1H-pyrazole-3-carboxylate, 2-aminopyridine, and isocyanide with ZnO nanoparticles as catalyst (Figure 2.8). This methodology has many advantages such as it is eco-friendly, can tolerate wide functional groups, is used as recyclable catalyst, and has a short reaction time [25].

2.4.1.4   Synthesis of Quinoxalines

Sadeghi et al. reported a simple protocol for the synthesis of quinoxaline derivatives by using 1,2-diphenyl diamine and 1,2-dicarbonyl with ZnO nanoparticles as solid acid catalyst under solvent-free conditions at room temperature (Figure 2.9) [26]. This simple methodology offers several advantages including a mild reaction condition, a simple work-up, opportunities for scale-up, and improved yields. The reaction proceeds through the coordination of Lewis acid sites of ZnO (Zn^{2+}) with the oxygen of the carbonyl group, and hence the reactivity of the carbonyl group increases. Then a nucleophilic attack to the activated carbonyl groups proceeds the reaction forward.

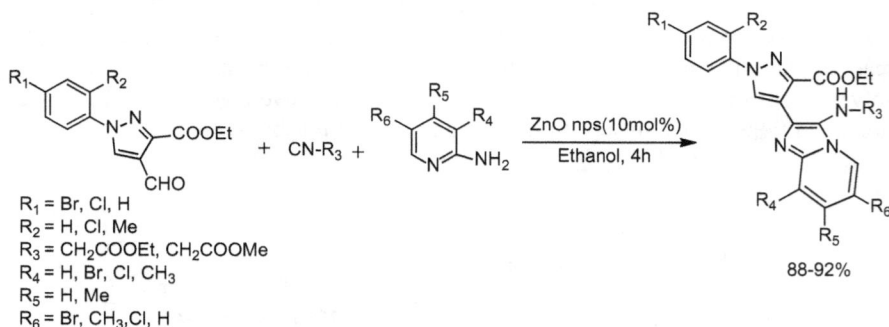

R_1 = Br, Cl, H
R_2 = H, Cl, Me
R_3 = CH_2COOEt, CH_2COOMe
R_4 = H, Br, Cl, CH_3
R_5 = H, Me
R_6 = Br, CH_3, Cl, H

FIGURE 2.8   Synthesis of imidazo-fused polyheterocycles.

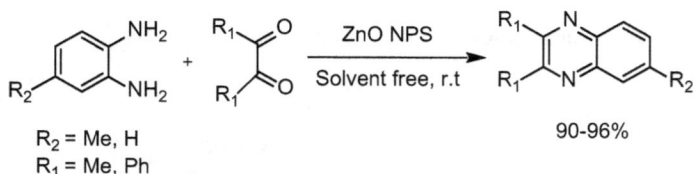

R_2 = Me, H
R_1 = Me, Ph

FIGURE 2.9   Synthesis of quinoxalines.

2.4.2 Synthesis of Oxygen-containing Heterocycles

2.4.2.1 Synthesis of Furan Derivatives

Furanones are five-membered heterocyclic compounds having lactone rings in their structures, and they are present as the core structure of many bioactive natural products and drugs. Tekale and co-workers had established an effective method for synthesis of 3,4,5-trisubstituted furan-2(5H)-ones from aldehyde, amines, and dimethylacetylenedicarboxylate (DMAD) using nano-ZnO as reusable heterogeneous catalyst. The methodology is preferable due to its simplicity, short reaction time, high yield, and cheap heterogeneous catalyst (Figure 2.10) [27]. The first step in the mechanism involves the formation of enamine from amine and DMAD. ZnO polarizes the carbonyl group to generate a polarized adduct, which interacts with enamine, cyclizes, and eliminates the methanol molecule.

2.4.2.2 Synthesis of coumarin

A large number of substituted coumarin derivatives were synthesized from acetonitrile, ethyl acetoacetate or ethyl benzoyl acetate, and phenol derivatives by using the combination of organo- and nanocrystalline ZnO catalyst [28]. Usage of triethanolamine has notable influence on the ZnO morphology and thus reactivity of the catalyst. The devised method is a simple and benign method for the synthesis of substituted coumarins using an environmentally benign combination of organo- and nano-cocatalyst within a short reaction time. Another added advantage is the absence of the formation of chromones as the side products. Furthermore, the lack of purification procedure makes this a valuable alternative to prevailing methods.

Kumar et al. reported the use of ZnO nanoparticles for the synthesis of coumarins in moderate to excellent yield from derivatives of salicylaldehyde and 1,3-dicarbonyl compounds under microwave irradiation [29]. The catalyst can be reused and recycled without losing activity and selectivity.

2.4.2.3 Synthesis of Xanthenes

Ghomi et al. reported the synthesis of xanthene derivatives via a multicomponent reaction of aldehyde, dimedone, and 2-napthol using ZnO nanoparticles under solvent-free conditions. The methodology is effective as the catalyst can be recovered easily, is eco-friendly, and needs low catalyst loading to obtain corresponding

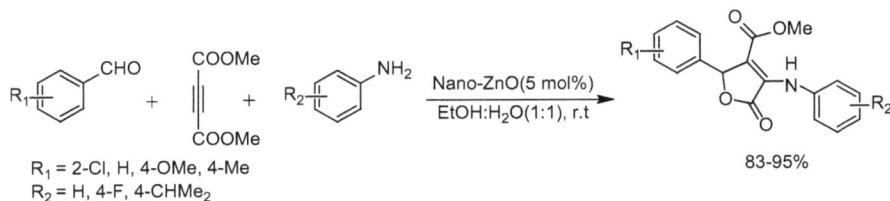

R_1 = 2-Cl, H, 4-OMe, 4-Me
R_2 = H, 4-F, 4-CHMe$_2$

83-95%

FIGURE 2.10 Synthesis of 3,4,5-trisubstituted furan-2(5H)-ones.

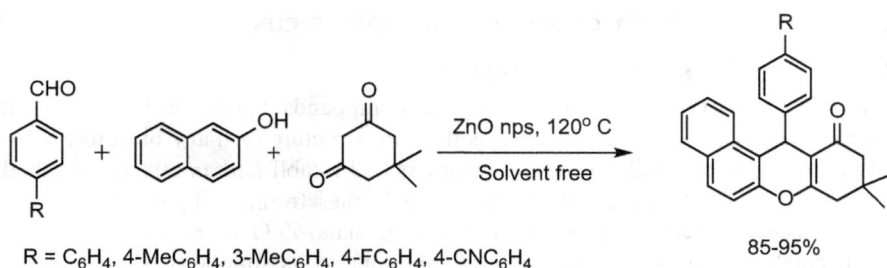

R = C$_6$H$_4$, 4-MeC$_6$H$_4$, 3-MeC$_6$H$_4$, 4-FC$_6$H$_4$, 4-CNC$_6$H$_4$

85–95%

FIGURE 2.11   Synthesis of xanthene derivatives.

xanthenes in high yield (Figure 2.11) [30]. It was observed that on increasing the catalyst amount from 5 to 10 mol%, a better yield was obtained, but a further increase of the molar amount of catalyst to 15 mol% did not increase the product yield.

2.4.3   SYNTHESIS OF OTHER HETEROCYCLES

2.4.3.1   Synthesis of 6-amino-5-cyano-pyrano[2,3-c] Pyrazoles

Tekale et al. described an efficient, simple methodology for the synthesis of substituted pyranopyrazoles in excellent yield via a four-component coupling reaction of aromatic aldehyde, malononitrile, ethyl acetoacetate, and hydrazine hydrate in an aqueous medium using ZnO nanoparticles as a recyclable heterogeneous catalyst (Figure 2.12). The use of water and ZnO nanoparticles makes this protocol environmentally benign [31]. Operational simplicity, catalyst recyclability, atom economical, and environmentally benign nature all make this an attractive protocol.

2.4.3.2   Synthesis of Structurally Diverse Pyridine Derivatives

Siddiqui et al. reported a cost-effective and facile method for synthesis of pyridine derivatives via three-component reaction of β-enaminones, ammonium acetate, and various active methylene compounds followed by Michael addition, cyclodehydration, and elimination sequence under solvent-free conditions (Figure 2.13) [32]. The group has tested different solvent systems and catalysts for pyridine derivatives, but among them, ZnO NPs and solvent-free conditions were found to be effective in terms of yield and effectiveness. The catalyst was recyclable up to six catalytic cycles without a significant loss in the catalytic activity.

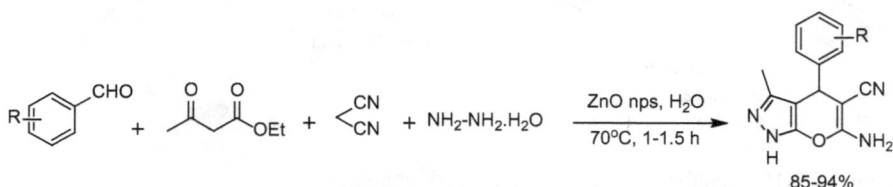

85–94%

FIGURE 2.12   Synthesis of pyranopyrazoles.

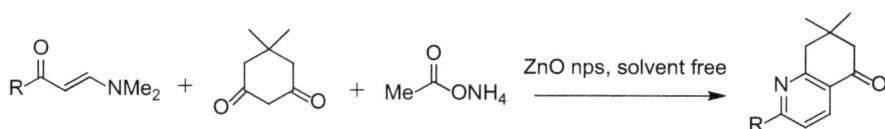

FIGURE 2.13 Synthesis of pyridine derivatives.

2.4.3.3 Synthesis of 2-thioxo-1,3-oxazoles Derivatives

Haerizade et al. have performed the reaction of ammonium thiocyanate, acid chloride, and phenacyl bromide or its derivatives in the presence of ZnO NPs catalyst and N-methylimidazole to synthesize 2-thioxo-1,3-oxazoles in 95% isolated yield (Figure 2.14) [33]. The solvent-free conditions and the simplicity of the procedure make it an interesting alternative to the more complex multistep approaches.

2.4.3.4 Synthesis of Bis-isoquinolinones

The isoquinoline nucleus consists of an aromatic ring fused with pyridine and is isolated along with other biologically important plant alkaloids. Krishnakumar and co-workers represented an effective ZnO nanoparticle-mediated reaction for the synthesis of bis-isoquinolines. Synthesis of bis-isoquinolines involved two steps – firstly, homophthalic acid reacted with acid chloride to give isocoumarins, which on further condensation with 1,7-heptadiamine afforded bis-isoquinoline in good yields (Figure 2.15) [34]. It was seen that rate of reaction increases in the presence of EDG

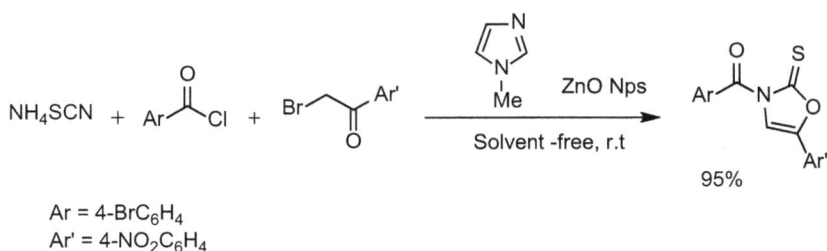

$Ar = 4\text{-}BrC_6H_4$
$Ar' = 4\text{-}NO_2C_6H_4$

FIGURE 2.14 Synthesis of 2-thioxo-1,3-oxazoles.

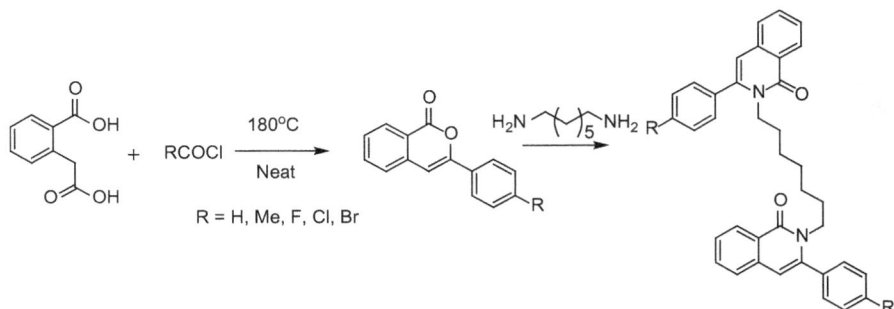

FIGURE 2.15 Synthesis of bis-isoquinolinones.

as compared with EWG. The methodology utilizes mild reaction conditions, which makes it proficient in comparison to other techniques.

2.5 TITANIUM OXIDE NANOPARTICLE-CATALYZED ORGANIC REACTIONS

Titanium oxide (TiO_2) is a naturally occurring oxide of titanium found in nature. Nowadays, TiO_2 nanoparticles are utilized in many organic reactions and for the synthesis of bioactive heterocyclic compounds.

2.5.1 STRECKER REACTION CATALYZED BY NANO-TiO_2 P25

Degussa TiO_2 (P25) is a well-known and widely investigated catalyst and photocatalyst. Strecker reaction is one of the most effective ways for the synthesis of bifunctional α-aminonitrile compounds. It provides a variety of amino acid precursors to both industries and laboratories. A simple and effective protocol was formulated for the synthesis of α-aminonitriles from aldehyde, trimethylsilyl cyanide, and amine in the presence of nano TiO_2 P25 reusable catalyst at room temperature in higher yields [35]. It was observed that nano-TiO_2 P25 was found to be a highly active catalyst for the synthesis of α-aminonitriles.

2.5.2 SYNTHESIS OF 2,3-DISUBSTITUTED DIHYRDOQUINAZOLIN-4(1H)-ONES

TiO_2 nanoparticles were utilized for the synthesis of 2,3-disubstituted dihydroquinazolin-4(1H)-ones under solvent-free conditions. It was noticed that increasing catalyst loading from 2 to 5 mol% improved the yield of the product from 49% to 91%; hence, the catalyst concentration played an important role in determining the product yield (Figure 2.16) [36]. Different solvents were screened for the reaction, and it was found that TiO_2 NPs synthesized quinazolinones at a faster rate and in good yield under solvent-free conditions.

2.5.3 FRIEDEL–CRAFTS ALKYLATION OF INDOLES WITH EPOXIDES CATALYZED BY NANOCRYSTALLINE TITANIUM(IV) OXIDE

Kantam et al. reported on the use of a recyclable, nanocrystalline TiO_2 catalyst for the Friedel–Crafts alkylation of indoles. The reaction of indole with styrene oxide afforded 2-(3-indolyl)-2-phenylethanol in 64% yield in the presence of DCM solvent and nano TiO_2. It was observed that the indoles having EDG enhanced the reaction rate, hence giving alcohols in moderate to good yields [37]. High regioselectivity was

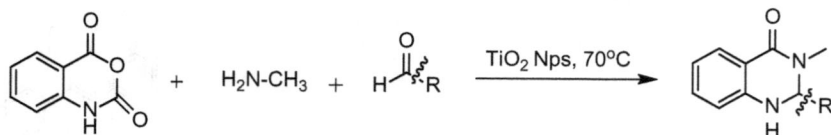

FIGURE 2.16 Synthesis of 2,3-disubstituted dihyrdoquinazolin-4(1H)-ones.

observed with TiO$_2$ catalyst. Nano-TiO$_2$ showed higher activity than other catalysts due to its distinct shape, size, and high surface area.

2.5.4 SYNTHESIS OF POLYSUBSTITUTED PYRROLIDINONES

Mukhopadhyay et al. were the first to report the micellar method as a reaction template by TiO$_2$ nanoparticles at room temperature in an aqueous CTAB (cetyl-trimethylammonium bromide) solution for the synthesis of a variety of 3-hydroxy-2-pyrrolidinones (Figure 2.17). The utilization of water as a reaction medium makes this protocol economical and eco-friendly [38]. It was analyzed from the DLS results that in the absence of CTAB in the aqueous medium, TiO$_2$ NPs aggregated themselves to form large particles, and no micelles were formed. The catalytic activity of other catalysts was also tested, and it was observed that 10 mol% of TiO$_2$ in 0.8 mm of aqueous CTAB solution catalyzed the reaction effectively to afford the product an excellent yield.

2.5.5 SYNTHESIS OF 2-ARYLBENZIMIDAZOLES AND 2-ARYLBENZOTHIAZOLE

Derivatives of benzimidazoles and benzothiazoles were prepared by using the H$_2$O$_2$/TiO$_2$ nanoparticle system from 12-phenylenediamines and 2-aminothiophenol with a wide range of aryl aldehydes as the starting material (Figure 2.18) [39]. Both EWG and EDG aldehydes gave corresponding benzimidazoles in good yields. The methodology has advantages such as it involves a simple workup procedure, has excellent chemoselectivity, is cost-effective, and is environmentally benign.

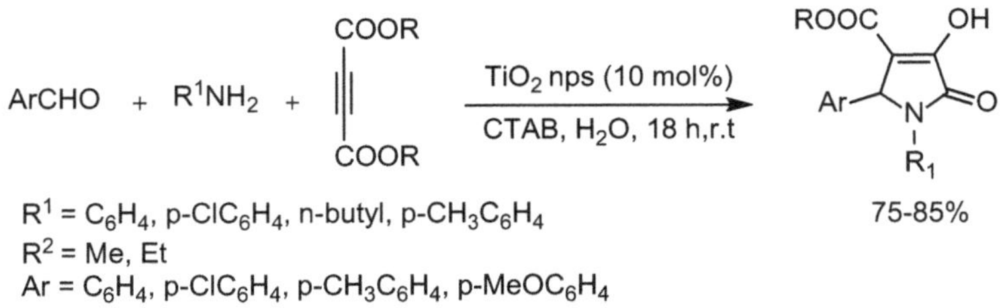

R^1 = C$_6$H$_4$, p-ClC$_6$H$_4$, n-butyl, p-CH$_3$C$_6$H$_4$
R^2 = Me, Et
Ar = C$_6$H$_4$, p-ClC$_6$H$_4$, p-CH$_3$C$_6$H$_4$, p-MeOC$_6$H$_4$

FIGURE 2.17 Synthesis of 3-hydroxy-2-pyrrolidinones.

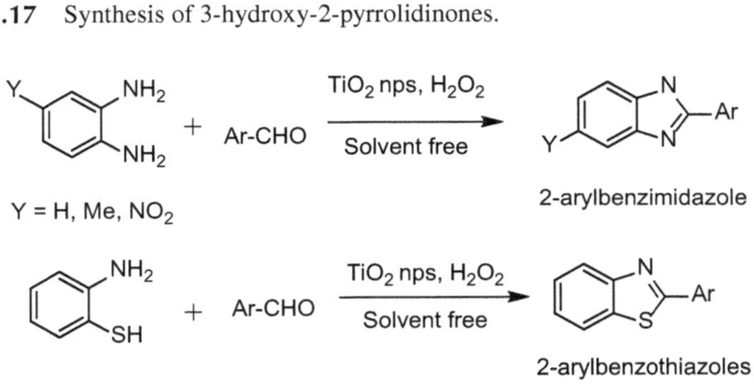

FIGURE 2.18 Synthesis of 2-arylbenzimidazoles and 2-arylbenzothiazole.

2.6 NICKEL OXIDE NANOPARTICLES IN ORGANIC TRANSFORMATIONS

Nickel oxide (NiO) nanoparticles act as nanocatalysts for many organic reactions because of their large surface-to-volume ratio and tunable structural morphology. NiO NPs have been used as an effective catalyst for the synthesis of various organic compounds, which are discussed below.

2.6.1 Synthesis of Spiro and Condensed Indole Derivatives

Saroj et al. reported on the synthesis of complex and various spiro and condensed indole derivatives involving Knoevenagel condensation followed by Michael addition under microwave irradiation using NiO nanoparticles. NiO nanoparticles can be recovered and reused easily without any significant loss in activity (Figure 2.19) [40]. Because of the high surface area and low coordination sites of NiO catalyst, it is found to be more reactive, leading to a high yield of product.

2.6.2 Amidoalkyl Naphthol Derivatives Synthesis

Juneja et al. were able to synthesize NiO nanoparticles by the sol–gel method, which were further utilized as a catalyst for the synthesis of amidoalkyl naphthol derivatives by condensation of aromatic aldehyde, β-napthol, and amide for the first time. The opted methodology was simple and compatible with different functional groups; the eco-friendly and corresponding product was obtained with high yields (Figure 2.20) [41]. The catalyst displayed excellent recyclability and reusability up to four times without further treatment. On catalyst loading from 3 to 12 mol%, the yield was increased from 32% to 94%, but no further increment was seen in yield when the concentration was increased from 12 to 18 mol%.

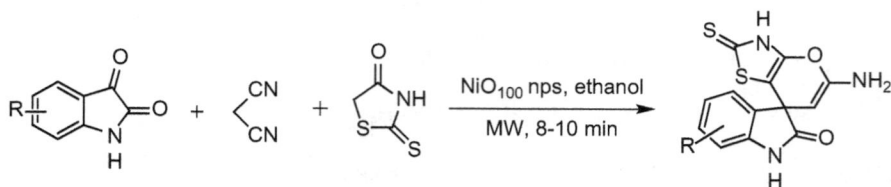

FIGURE 2.19 Synthesis of spiro and condensed indole derivatives.

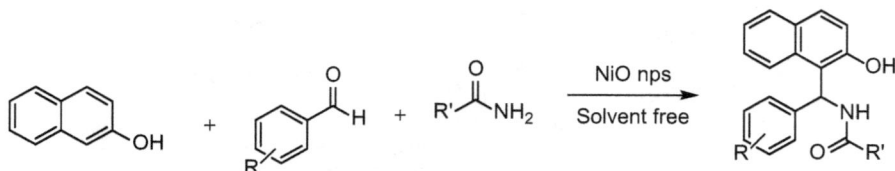

R' = CH$_3$, NH$_2$
R = Me, 4-NH$_3$, 3-NO$_2$, 4-NO2

FIGURE 2.20 Synthesis of amidoalkyl naphthol derivatives.

2.6.3 Synthesis of Diindolyl Oxindole Derivatives in Aqueous Medium

A green and effective protocol for the synthesis of diindolyl oxindole, a crucial class of bioactive compounds, was reported by Nasseri et al. Treatment of indole with isatin in the presence of water using NiO nanoparticles afforded oxindole derivatives in excellent yield (Figure 2.21) [42]. The mechanism involved is interesting as the first step requires the formation of activated isatin, followed by reaction with indole and elimination, which further produces an intermediate that reacts with a second indole molecule to afford an oxindole derivative.

Some other transition metal oxide nanoparticles catalyzing organic reactions are summarized in Table 2.2.

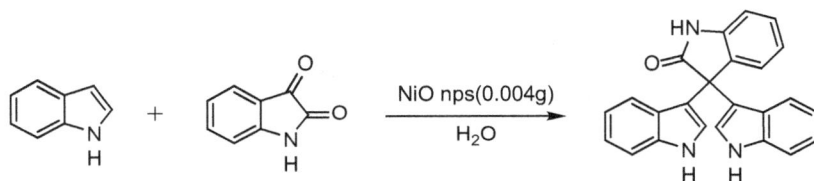

FIGURE 2.21   Synthesis of 3,3-diindolyloxindole.

TABLE 2.2
Transition Metal Oxide Nanoparticle-catalyzed Organic Reactions

S. No.	Organic Reactions Catalyzed by Transition Metal Oxides	Reference
1	Synthesis of benzimidazole by Pt-coated TiO_2 NPs	[43]
2	Synthesis of trifluromethyl-1,2,4-oxadiazoles using TiO_2 NPs	[44]
3	Synthesis of 2-oxo-2,5-dihydropyrroles catalyzed by TiO_2 NPs	[45]
4	Synthesis of 9-aryl-1,8-dioxo-octahydroxanthenes by TiO_2 NPs	[46]
5	Synthesis of α-hydroxyphosphonates using TiO_2 NPs under MW	[47]
6	Synthesis of 2,3-disubstituted quinazolin-4(1H)-ones by RuO_2 NPs	[48]
7	Epoxidation of alkenes using MnO_2 NPs	[49]
8	Hydrolysis of ethyl acetate catalyzed by Mn_3O_4 NPs	[50]
9	Epoxidation of styrene using Fe_3O_4 NPs	[51]
10	Synthesis of triazolo, tetrazolo[1,5-α]pyrimidine derivatives using CeO_2 NPs	[52]
11	Synthesis of 2,4,6-triaryl pyridines by using Fe_3O_4 NPs	[53]
12	Synthesis of 2-oxocoumarin-3-carboxamides by Fe_3O_4 NPs	[54]
13	Azo coupling of aryl amines in presence of RuO_2/Cu_2O NPs	[55]
14	Synthesis of polyhydroquinoline by CuO NPs	[56]
15	Synthesis of polyhydroquinoline using CeO_2 NPs	[57]
16	Synthesis of 1,8-dioxo-octahydroxanthenes using CeO_2 NPs	[58]
17	Synthesis of xanthene derivatives by Fe_3O_4 NPs	[59]
18	Synthesis of 3,4-dihydropyrimidine-2(1H)-ones using Fe_3O_4 NPs	[60]
19	Synthesis of N-monosubstitued ureas using CeO_2 NPs	[61]
20	Synthesis of xanthene derivatives in presence of ZnO NPs	[62]
21	Synthesis of 5-arylidine barbituric acid by CuO NPs	[63]

2.7   TRANSITION METAL FERRITES IN ORGANIC REACTIONS

Ferrites are chemical compounds in powder form possessing ferrimagnetic properties, made from iron oxides as their fundamental component. They can be categorized into hexagonal ($MFe_{12}O_{19}$), garnet ($M_3Fe_5O_{12}$), and spinel (MFe_2O_4) form, where M stands for one or more divalent transition metals (Zn, Mn, Co, Ni, Cu). Transition metal ferrites are used in catalysis, as they can be easily recovered with the help of a magnet after the reaction completion. Various types of transition metal ferrites, along with their applications, are described below.

2.7.1   COBALT FERRITE

Cobalt ferrite ($CoFe_2O_4$) possesses an inverse spinel structure, where divalent Co^{2+} occupies the octahedral sites while Fe^{3+} occupies both tetrahedral and octahedral sites equally [64, 65]. Several techniques like microemulsion, sol–gel method, hydrothermal method, co-precipitation method, electrochemical method, and solvothermal method can be used to synthesize $CoFe_2O_4$ [66–71]. They exhibit properties like high coercivity, good mechanical strength, moderate saturation magnetization, super magnetism, and high chemical stability. For catalysis, they should have narrow size distribution and high magnetization value [72]. Because of these features, cobalt ferrites are utilized in recording devices, sensors, magnetic cards, magnetic drug delivery, and catalysis. The saturation magnetization (MS) of cobalt ferrite NPs is less in comparison to bulk materials, and it decreases with a decrease in size.

2.7.1.1   Oxidation Reactions

2.7.1.1.1   Oxidation of Alcohol

Ramzani et al. reported an effective method for oxidation of benzylic and aliphatic alcohols in the presence of potassium hydrogen monopersulfate and $CoFe_2O_4$ nanoparticles as a catalyst to afford corresponding carbonyl compounds in good yield. The catalyst was effective without loss in its activity even after six catalytic cycles [73]. It was observed that EWG decreased the reaction rate while EDG on benzene enhanced the rate of reaction. Oxidation with potassium hydrogen monopersulfate occurred in a short reaction time and at a low temperature compared to O_2 and tert-Butyl hydroperoxide (TBHP). The use of inexpensive and non-toxic materials, short reaction time, mild reaction time, and catalyst reusability make this an attractive protocol.

2.7.1.1.2   Oxidation of Hydrocarbon

Xia and co-workers were able to synthesize spinel cobalt ferrite nanoparticles by sol–gel auto combustion method and utilized them for the oxidation of cyclohexane in the presence of oxygen as an oxidizing agent under solvent-free conditions. The product was obtained in high yield with 288 turnover numbers and 92.4% selectivity for cyclohexanone and cyclohexanol under 1.6 MPa of initial oxygen pressure at 418 K after 6.0 h of reaction. [74]. After completion of the reaction, the catalyst can be easily separated by using an external magnet; although a slight loss in its activity was observed, the selectivity remained intact.

2.7.1.2 Coupling Reactions

2.7.1.2.1 Carbon–Oxygen Bond

A new methodology for C–O coupling reaction was reported between several aromatic alcohols and aryl halides by using cobalt ferrite nanoparticles in the presence of DMF solvent and K_2CO_3 base. Catalyst can be recycled by an external magnetic field [75]. The effect of different polar protic and aprotic solvents was tested and studied with K_2CO_3 base in the presence of $CoFe_2O_4$ NPs, but only DMF solvent was found to be effective for the C–O coupling reaction. It was noticed that the presence of EWG and sterically hindered groups on phenol retarded the reaction rate, but EDG enhanced the reaction rate.

2.7.1.2.2 Carbon–Carbon Bond (Suzuki Coupling)

The carbon–carbon coupling reaction is one of the most widely used reactions in organic synthesis as it is utilized in the production of various industrial, pharmaceutical, and biological products. The catalytic activity of cobalt ferrite nanoparticles was checked on the Suzuki coupling reaction of several aryl bromides or iodides with aryl boronic acid in refluxing ethanol and sodium carbonate. Many symmetric and asymmetric biaryl derivatives were prepared in good yield [76]. The catalyst can be recovered by using an external magnet and reused after the reaction completion.

2.7.1.2.3 Carbon–Sulfur Bond

Zhou et al. reported the C–S coupling of thiols with aromatic iodides using cobalt ferrite NPs in the presence of DMF and K_2CO_3. Treatment of 5-methyl-1,3,4-thiadiazol-2-thiol with iodobenzene afforded the desired product with 83% isolated yield (Figure 2.22) [77]. With no change in magnetic properties, the morphology of catalyst was observed even after five consecutive catalytic cycles, and hence it displayed good catalytic activity.

2.7.1.3 Ring-opening Reaction of Epoxides

Zhang et al. synthesized $CoFe_2O_4$-supported phosphomolybdate nanoparticles ($CoFe_2O_4@SiO_2$-$PrNH_2$-PMo) and used it for the ring-opening reaction of different epoxides with ethereal H_2O_2 at room temperature to get β-hydroxy hydroperoxides in good yield [78]. The benefits exhibited by the catalytic system are easy separation

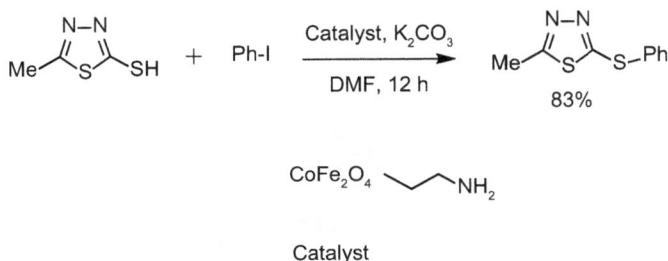

FIGURE 2.22 Coupling reaction between thiols and aromatic iodide.

and preparation, reusability, high catalytic activity, and stability, which make this protocol efficient.

2.7.1.4   Knoevenagel Condensation

Spinel cobalt ferrite nanoparticles were synthesized using both sonochemical and co-precipitation methods without any capping agent and surfactant. A reusable catalyst was employed for the Knoevenagel condensation reaction of aldehydes with ethyl cyanoacetate in aqueous ethanol under mild reaction conditions [79]. The excellent yield of desired Knoevenagel products was obtained in a short period of time with 5 mol% of the catalyst. It has been found that the NPs showed stability in both ethanol and aqueous solvent systems and can be stored for months without any stabilizer; no agglomeration was observed.

2.7.1.5   Ritter Reaction

Ritter reaction is one of the important reactions used for the preparation of amides [80]. Magnetic $CoFe_2O_4$ nanoparticle-immobilized diamine-N-sulfamic acid ($CoFe_2O_4@SiO_2$–DASA) was prepared and used as a catalyst for the synthesis of amides under solvent-free conditions. Catalyst can be recovered and reused easily without any loss in its activity. [81]. The opted methodology was easy, non-toxic in terms of solvent, eco-friendly, and provided the corresponding product in higher yields.

2.7.1.6   Synthesis of 5-hydroxymethylfurfural

Inverse spinel cobalt ferrite nanoparticles were modified with structurally different dicarboxylic acids. The modified nanoparticle was able to catalyze fructose to produce HMF under solvent-free conditions (Figure 2.23). It has been observed that the flexibly modified ferrite showed good catalytic activity in comparison with rigid ligand-modified ferrites [82]. It was noticed that the binding mode of the ligand and its orientation with a surface modifier were important for determining the yields and selectivity.

2.7.2   Copper Ferrite Nanoparticles

Copper ferrite ($CuFe_2O_4$) is an important ferrite, which comes under the category of spinel ferrites. Widely used methods for synthesizing copper ferrite nanoparticles are co-precipitation, sol–gel, hydrothermal method, and solid reaction process, but solid reactions take a long time. They have the ability to modify physical properties when exposed to different environments [83]. $CuFe_2O_4$ nanoparticles are less expensive, recyclable, require mild reaction conditions, and hence are employed as heterogeneous catalysts in various organic transformations.

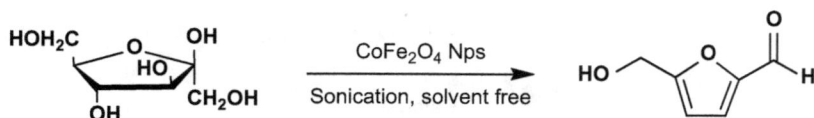

FIGURE 2.23   Synthesis of HMF.

2.7.2.1 Synthesis of 1,4-dihydropyridines

1,4-Dihydropyridines are the class of drugs that helps in calcium channel modulation and is common to many bioactive compounds. A simple and effective protocol was developed by Murthy et al. for the synthesis of 1,4-dihydropyridines (Hantzsch esters). Treatment of substituted aromatic aldehyde, ethyl acetoacetate, and ammonium acetate in ethanol in the presence of copper ferrite nanoparticles at room temperature afforded corresponding 1,4-dihydropyridine [84]. The methodology provides several advantages like low catalyst loading, short reaction times, high product yields, and easy catalyst separation.

2.7.2.2 Synthesis of α-aminonitriles

α-Aminonitriles are the main intermediates for synthesizing amino acids, amides, nitrogen-containing heterocycles, and diamide. An efficient method was prepared for the synthesis of α-aminonitriles by various aldehydes, amines, and trimethylsilyl cyanides in water using copper ferrite nanoparticles at room temperature [85]. The mechanism involves activating aldehyde by the acidic proton of copper ferrite nanoparticles which further reacts with amines to form the intermediate and finally leads to the formation of α-aminonitriles.

2.7.2.3 Ullmann C–O Coupling Reaction

Wang et al. reported the use of copper ferrite nanoparticles for the Ullmann type C–O coupling reaction between phenols and aryl halides. The reaction occurred smoothly even in the presence of sensitive substituents (CH_3CO, NH_2, CN) to afford the product in high yield [86] without protecting other functional groups. The catalyst is stable toward the air, could be made easily and recycled with the help of an external magnet.

2.7.2.4 Synthesis of β,γ-unsaturated Ketones

The reaction of different acid chlorides with allyl halides in THF solvent using copper ferrite nanoparticles at room temperature to give β,γ-unsaturated ketones in good yield was reported by Murthy et al. The methodology is free from any additive or co-catalyst [87]. The remarkable features of this method are less reaction time, mild reaction conditions, no isomerization during the reaction, and heterogeneous reusable catalyst.

2.7.2.5 Direct C–H Amination of Benzothiazoles

Benzothiazole acts as an eminent bioactive molecular entity used as an anti-HIV agent, H3-receptor ligand, antibacterial compound, and PPAR agonists. An efficient and simple methodology was reported for synthesizing 2-N-substituted benzothiazoles using copper ferrite nanoparticles in the presence of a Cs_2CO_3 base with nitrogen nucleophiles [88]. The catalyst can be recycled and reused without any significant loss in its activity. This methodology is effective and useful, as many five- and six-membered heteroaromatic compounds are seen in many biologically active compounds and organic molecules.

2.7.2.6   Synthesis of 9-substituted Aryl-1,8-dioxo-octahydroxanthenes

Xanthene and its derivatives are the crucial class of organic molecules as they possess extensive biological and pharmaceutical properties. Murthy et al. reported on the reaction between dimedone with different aromatic aldehydes using copper ferrite nanoparticles to afford 9-substituted aryl-1,8-dioxo-octahydroxanthenes in excellent yield under mild reaction conditions. It was observed that aldehydes having EWG needed less reaction time in comparison with aldehydes having EDG because EWG accelerates the formation of carbocation at the carbonyl carbon of aldehyde [89]. Different mol% of catalyst was evaluated for the reaction, and it was found that 15 mol% of catalyst was efficient to produce the desired product in high yield in less reaction time.

Some other organic reactions catalyzed by copper ferrite nanoparticles are discussed in Table 2.3.

2.7.3   NICKEL FERRITE NANOPARTICLES

Nickel ferrite nanoparticle ($NiFe_2O_4$) is an inverse spinel that consists of 8 $NiFe_2O_4$ molecules in a unit cell. Half of the Fe^{3+} occupies the tetrahedral sites (A-sites), while others occupy the octahedral sites (B-sites). They show ferromagnetism which arises from the magnetic moment of antiparallel spins between Fe^{3+} in A site and Ni^{2+} in B site [99]. The method for their synthesis is the same as that of other ferrites, but pulsed laser deposition is a widely used method. $NiFe_2O_4$ displays a super magnetic nature and hence can be utilized as catalysts, gas sensors, magnetic fluids, and magnetic storage systems.

2.7.3.1   C–O Bond Formation

A novel methodology for C–O coupling reaction was reported by Moghaddam et al. between several aromatic alcohols and aryl halides by using nickel ferrite nanoparticles in the presence of DMF solvent and K_2CO_3 base. Catalyst can be recycled with the help of an external magnetic field, and reaction occurs in a short duration of

---

TABLE 2.3
Nano-sized Copper Ferrite-catalyzed Organic Reactions

S. No.	Organic reactions	References
1	Synthesis of propargylamine by $CuFe_2O_4$ NPs	[90]
2	Synthesis of sulfones using $CuFe_2O_4$ NPs	[91]
3	N-Arylation of pyrroles with aryl halides in the presence of $CuFe_2O_4$ NPs	[92]
4	Synthesis of 2-substituted benzoxazole using $CuFe_2O_4$ NPs	[93]
5	Synthesis of phenyl acetone by $CuFe_2O_4$ NPs	[94]
6	Synthesis of benzonitrile derivatives by Pt-coated $CuFe_2O_4$ NPs	[95]
7	O-Arylation reaction catalyzed by $CuFe_2O_4$ NPs	[96]
8	Sonogashira reaction catalyzed by Pd-$CuFe_2O_4$ NPs coated at SiO_2	[97]
9	Thioetherification reaction catalyzed by $CuFe_2O_4$ NPs	[98]

TABLE 2.4
Transition Metal Ferrite NP-catalyzed Organic Reactions

S. No.	Organic Reactions	Reference
1	Synthesis of chalcone derivatives using $ZnFe_2O_4$ NPs	[100]
2	Synthesis of nopol by $ZnFe_2O_4$ NPs	[101]
3	Synthesis of benzonitrile by $NiFe_2O_4$ NPs	[102]
4	Synthesis of disulfides by $NiFe_2O_4$ NPs	[103]
5	Synthesis of spirooxindoles using $MnFe_2O_4$ NPs	[104]

time [75]. Different solvents and bases were employed for the reaction, and it was found that both K_2CO_3 and Cs_2CO_3 gave the product an excellent yield; due to the lower cost of K_2CO_3 it was favored. Both DMF and K_2CO_3 afforded the product in high yield. More organic reactions catalyzed by nickel nanoparticles are mentioned in Table 2.4 (entries 3 and 4). Apart from applications in organic synthesis Nickel ferrite nanoparticles have applications in industries also like water gas shift reaction.

2.7.4 ZINC FERRITE NANOPARTICLES

Nano zinc ferrites are crucial among other ferrites due to some properties like low-saturation magnetization and high resistivity, and hence it is utilized in data recording media, sensors, and adsorption processes [105–107].

Balint et al. were able to prepare neodymium (Nd)-substituted ferrites and utilized them to check the catalytic efficiency for the first time for the oxidative conversion of methane in oxidative conditions (methane combustion) and reductive conditions (oxidative coupling of methane). It was observed that Nd-substituted ferrites showed excellent catalytic efficiency for methane combustion while $ZnFe_2O_4$ and $ZnNd_2O_4$ were effective for oxidative coupling of methane. Hence, the redox properties of simple and substituted ferrites and the nature of substitutional elements were crucial parameters in the determination of catalytic properties [108]. The catalytic activity was found to be dependent on the structure of oxide to the specific defects generated by substitution and composition of the reaction mixture. Other examples of organic reactions catalyzed by zinc ferrite nanoparticles are quoted in Table 2.4 (entries 1 and 2).

2.8 MIXED METAL OXIDES NANOPARTICLES

Mixed metal oxide (MMO) nanoparticles are also known as heterometal oxide nanoparticles. MMO is a widely utilized class of solid catalysts, either as supports or active phases. They have various roles in different fields of chemistry, but the major one is found in catalysis due to the distinctive magnetic and electronic properties acquired after a combination of two metals in the oxide matrix [109, 110]. $BaTiO_3$, $LiNbO_3$, and $KTaO_3$ possess high dielectric, Ferro, or pyroelectric properties, hence have many applications in industries. MMO can either be crystalline or amorphous in nature.

MMO nanoparticles can be synthesized by hydrothermal, sol–gel, co-precipitation, wet impregnation, and mechanochemical methods [111–115], but sol–gel is the most widely used method. The degree of dispersion in the chemical route is totally dependent on preparation technique and provided synthetic conditions. Some mixed metal oxides are good in terms of catalytic activity as compared to component oxides in various reactions, the reason behind this is increasing active acidic or basic sites and surface area that decreases the reaction time, and hence yield of the reaction is increased [116–118]. The best example of increased catalytic activity is observed in mixed aerogel of titania/zirconia, in which the acidic strength and number of active sites are increased compared with pure titania and zirconia. It suggests that combining two metal oxides is an efficient method to bring two oxides close [119]. MMO nanoparticles properties are size dependent; physical and chemical properties are distinct and remarkable as compared to their corresponding bulk material due to the gradual shift from atomic or molecular form to condensed matter structure. This change is making MMO nanoparticles more reactive, as atoms are now more in activated form in comparison with bulk counterparts.

2.8.1 MgO-ZrO_2-MIXED METAL OXIDE NANOPARTICLE

Gawande et al. were able to synthesize the proficient heterogeneous nano magnesia-zirconia (MgO-ZrO_2) catalyst by ultra-dilution procedure. Prepared catalyst was reusable and had high surface area, and the catalyst was tested for the following reactions:

2.8.1.1 Cross Aldol Reaction

A crossed aldol condensation uses two different aldehydes and/or ketone reactants. Even though such reactions usually give a mixture of multiple condensation products, because there are two or more possible enolates and two different carbonyl electrophiles, it is the standard way to form carbon–carbon bonds in many reactions [120]. In most cases, an intelligent combination of carbonyls is taken: one of the carbonyls is usually an aromatic aldehyde devoid of any enolizable hydrogens, while the other having an acidic hydrogen acts as an enolate nucleophile. The reaction of an aromatic aldehyde with cyclic ketone using MgO-ZrO_2 NPs under solvent-free conditions at room temperature gives α–α'-bis-cycloalkanones an excellent yield [121]. It has been noticed that MgO-ZrO_2 NPs displayed good catalytic activity due to high surface area under mild reaction conditions. In this reaction the Lewis-acidic character of metal-oxide is used for activating the carbonyl toward nucleophilic addition reaction.

2.8.1.2 N-benzyloxycarbonylation of Amines

Benzyloxycarbonyl-protected amines are effective precursors for different natural and pharmaceutical products [122]. Several amines were examined for the benzyloxycarbonylation of amines by MgO-ZrO_2 catalyst under solvent-free conditions at 25 °C. All amines (aromatic, aliphatic, and heterocyclic) were able to give the corresponding product in high yield [121]. The catalyst was recyclable and had a high surface area which facilitates the catalytic activity.

2.8.1.3   Reduction of Aromatic Nitro Compounds

MgO-ZrO$_2$ NPs were tested for the catalytic transfer hydrogenation reaction of aromatic nitro compounds in the presence of KOH promoter, and isopropyl alcohol was used as a hydrogen source to give corresponding amines. It was observed in the absence of KOH and hydrogen source; the corresponding product was not achieved [121]. All aromatic nitro substituted compounds were suitable for the reaction, and no side products were noticed.

2.8.1.4   Synthesis of 1,5-benzodiazepines

1,5-Benzodiazepines are the biologically significant molecules, which are widely used as sedative, analgesic, and anti-depressive agents. It was synthesized conveniently by the treatment of cyclohexanone with orthophenylene diamine using MgO-ZrO$_2$ nanoparticles as catalyst under a solvent-free condition. It was observed that reaction proceeded well with aliphatic, cyclic, and aromatic ketones to get desired product in good yield [121]. The catalyst displayed good catalytic ability and can be recovered easily without any loss in activity.

2.8.1.5   Synthesis of Tinidazole

Tinidazole is an antimicrobial drug obtained by condensation of 2-methyl,5-nitro-imidazole and 2-ethyl-thio-ethanol using MoO$_3$/SiO$_2$ catalyst, which was prepared by the sol–gel technique to get 1-(2-ethyl-thio-ethanol)-2-methyl-5-nitro-imidazole followed by its oxidation using H$_2$O$_2$ oxidizing agent to get tinidazole (Figure 2.24). The prepared catalyst showed acidic strength and possessed a dual function as it was able to catalyze condensation and oxidation required during the synthesis of tinidazole. The use of MoO$_3$/SiO$_2$ catalyst made this process eco-friendly as usage of acetic acid in condensation, and ammonium molybdate or tungstic acid was avoided [123]. The catalyst can be recovered easily and reused five times without any loss in its activity and selectivity.

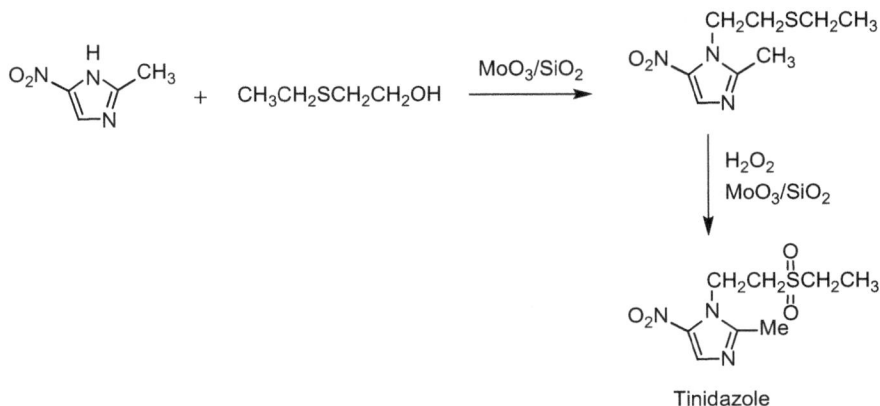

FIGURE 2.24   Synthesis of tinidazole.

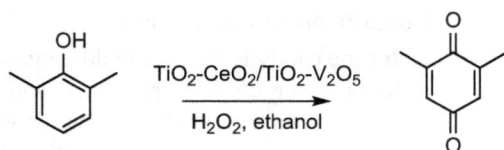

FIGURE 2.25 Oxidation of 2,6-dimethyl phenol.

2.8.1.6 Oxidation of 2,6-dimethyl Phenol

TiO_2 NPs in combination with other oxides have played a prominent role in the catalysis of important organic reactions. TiO_2-CeO_2 and TiO_2-V_2O_5 are important mixed oxides which were prepared by sol–gel method. These mixed-oxide gels were synthesized by loading different amount of TiO_2 and CeO_2 or V_2O_5 by using titanium isopropoxide and vanadyl acetyl acetone or hexahydrated cerium nitrate as precursors in ethanol as solvent. The prepared gels were used for the oxidation of 2,6-dimethyl phenol to obtain 2,6-dimethyl-p-benzoquinone using H_2O_2 as oxidant and ethanol as solvent (Figure 2.25) [124, 125].

2.8.1.7 Synthesis of 1,4-dihydropyridines

A very proficient methodology has been reported for the synthesis of Hantzsch 1,4-dihydropyridines through three component reaction of ethyl acetoacetate, ammonium acetate, and substituted aldehydes by using ZnO@SnO_2 MMO nanoparticles under solvent-free conditions. The protocol is effective in comparison with other reported reactions in terms of minimum catalyst loading, short reaction time, and no use of solvent, and high yield [126].

Apart from organic applications of MMO nanoparticles, they are widely used in industries and biological field due to their antibacterial and cytotoxic properties.

REFERENCES

1. Laurent, S., Forge, D., Port, M., Roch, A., Robic, C., Vander Elst, L., & Muller, R. N. (2010). Magnetic iron oxide nanoparticles: Synthesis, stabilization, vectorization, physicochemical characterizations, and biological applications. *Chemical Reviews*, *110(4)*, 2573–2574.

2. Astruc, D. (2020). Introduction: Nanoparticles in catalysis. *Chemical Reviews*, *120(2)*, 461–463.

3. Narayan, N., Meiyazhagan, A., & Vajtai, R. (2019). Metal nanoparticles as green catalysts. *Materials*, *12(21)*, 3602.

4. Kung, H. H., & Kung, M. C. (2007). Nanotechnology and heterogeneous catalysis. *Nanostructure Science and Technology*, 1–11. Springer: New York, NY.

5. Khan, I., Saeed, K., & Khan, I. (2017). Nanoparticles: Properties, applications and toxicities. *Arabian Journal of Chemistry*, *12(7)*, 908–931.

6. Loureiro, A., Azoia, N. G., Gomes, A. C., & Cavaco-Paulo, A. (2016). Albumin-based nanodevices as drug carriers. *Current Pharmaceutical Design*, *22(10)*, 1371–1390.

7. Todescato, F., Fortunati, I., Minotto, A., Signorini, R., Jasieniak, J., & Bozio, R. (2016). Engineering of semiconductor nanocrystals for light emitting applications. *Materials*, *9(8)*, 672.

8. Rogozea, E. A., Petcu, A. R., Olteanu, N. L., Lazar, C. A., Cadar, D., & Mihaly, M. (2017). Tandem adsorption-photodegradation activity induced by light on NiO-ZnO p–n couple modified silica nanomaterials. *Materials Science in Semiconductor Processing*, *57*, 1–11.

9. Nagarajan, P. K., Subramani, J., Suyambazhahan, S., & Sathyamurthy, R. (2014). Nanofluids for solar collector applications: A review. *Energy Procedia*, *61*, 2416–2434.

10. Bringmann, G., Gunther, C., Ochse, M., Schupp, O., & Tasler, S. (2001). In *Progress in the Chemistry of Organic Natural Products*. W. Herz, H. Falk, G. W. Kirby, & R. E. Moore (Eds.), (pp. 1–293). Springer: New York.

11. (a) Hegedus, L. S. (2002). In *Organometallics in Synthesis*. M. Schlosser (Ed.), (pp. 1123). J. Wiley & Sons: Chichester; (b) *Handbook of Organopalladium Chemistry for Organic Synthesis* (2002). E. Negishi (Ed.). Wiley-Interscience: New York; (c) Diederich, F., & Stang, P. J. (1998). *Metal-Catalyzed Cross-coupling Reactions*. Wiley-VCH: Weinheim, Germany; (d) Miyaura, N. (2002). *Cross-Coupling Reaction*. Springer: Berlin; (e) Meijere, A., & Diederich, F. (2004). *Metal-Catalyzed Cross-coupling Reactions*. Wiley-VCH: Weinheim, Germany; (f) Wang, D. P., Zhang, X. D., Liang, Y., & Li, J. H. (2006). *Chinese Journal of Organic Chemistry* 2006, 26, 19; (g) Stille, J. K. (1986). The palladium-catalyzed cross-coupling reactions of organotin reagents with organic electrophiles[new synthetic methods (58)]. *Angewandte Chemie International Edition in English*, *25(6)*, 508–524; (h) Hassan, J., Sévignon, M., Gozzi, C., Schulz, E., & Lemaire, M. (2002). Aryl–aryl bond formation one century after the discovery of the Ullmann reaction. *Chemical Reviews*, *102(5)*, 1359–1470; (i) Espinet, P., & Echavarren, A. M. (2004). The mechanisms of the Stille reaction. *Angewandte Chemie International Edition*, *43*, 4704–4734.

12. Li, J. H., Tang, B. X., Tao, L. M., Xie, Y. X., Liang, Y., Zhang, M. B. (2006). *Journal of Organic Chemistry*, *71*, 7488–7490.

13. Kidwai, M., Bhardwaj, S., & Poddar, R. (2010). C-Arylation reactions catalyzed by CuO-nanoparticles under ligand free conditions. *Beilstein Journal of Organic Chemistry*, *6*, 35.

14. Rout, L., Jammi, S., & Punniyamurthy, T. (2007). Novel CuO nanoparticle catalyzed C–N cross coupling of amines with iodobenzene. *Organic Letters*, *9(17)*, 3397–3399.

15. Li, J. H., Tang, B. X., Guo, S. M., & Zhang, M. B. (2008). N-Arylations of nitrogen-containing heterocycles with aryl and heteroaryl halides using a copper(I) oxide nanoparticle/1,10-phenanthroline catalytic system. *Synthesis*, *11*, 1707–1716.

16. Punniyamurthy, T., Jammi, S., Krishnamoorthy, S., Saha, P., Kundu, D., Sakthivel, S., Ali, M. A., & Paul, R. (2009). Reusable Cu_2O-nanoparticle-catalyzed amidation of aryl iodides. *Synlett*, *20*, 3323–3327.

17. Saha, P., Ramana, T., Purkait, N., Ali, M. A., Paul, R., & Punniyamurthy, T. (2009). Ligand-free copper-catalyzed synthesis of substituted benzimidazoles, 2-aminobenzimid-azoles, 2-aminobenzothiazoles, and benzoxazoles. *The Journal of Organic Chemistry*, *74(22)*, 8719–8725.

18. Sheldon, R. A. (2000). Atom efficiency and catalysis in organic synthesis. *Pure and Applied Chemistry*, *72(7)*, 1233–1246.

19. Jammi, S., Sakthivel, S., Rout, L., Mukherjee, T., Mandal, S., Mitra, R., Saha, P., & Punniyamurthy, T. (2009). CuO nanoparticles catalyzed C–N, C–O, and C–S cross-coupling reactions: Scope and mechanism. *The Journal of Organic Chemistry*, *74(5)*, 1971–1976.

20. Rout, L., Sen, T. K., & Punniyamurthy, T. (2007). Efficient CuO-nanoparticle-catalyzed C–S cross-coupling of thiols with iodobenzene. *Angewandte Chemie International Edition*, *46(29)*, 5583–5586.

21. Rao, K., Reddy, V., Swapna, K., & Kumar, A. (2009). Recyclable nano copper oxide catalyzed stereoselective synthesis of vinyl sulfides under ligand-free conditions. *Synlett*, *17*, 2783–2788.

22. Reddy, V. P., Kumar, A. V., Swapna, K., & Rao, K. R. (2009). Copper oxide nanoparticle-catalyzed coupling of diaryl diselenide with aryl halides under ligand-free conditions. *Organic Letters*, *11(4)*, 951–953.

23. Sabbaghan, M., & Ghalaei, A. (2014). Catalyst application of ZnO nanostructures in solvent free synthesis of polysubstituted pyrroles. *Journal of Molecular Liquids*, *193*, 116–122.

24. Alinezhad, H., Salehian, F., & Biparva, P. (2012). Synthesis of benzimidazole derivatives using heterogeneous ZnO nanoparticles. *Synthetic Communications*, *42(1)*, 102–108.

25. Swami, S., Devi, N., Agarwala, A., Singh, V., & Shrivastava, R. (2016). ZnO nanoparticles as reusable heterogeneous catalyst for efficient one pot three component synthesis of imidazo-fused polyheterocycles. *Tetrahedron Letters*, *57(12)*, 1346–1350.

26. Sadeghi, B., & Karimi, F. (2013). ZnO nanoparticles as an efficient and reusable catalyst for synthesis of quinoxalines under solvent free condition. *Iranian Journal of Catalysis*, *3*, 1–7.

27. Tekale, S. U., Kauthale, S. S., Pagore, V. P., Jadhav, V. B., & Pawar, R. P. (2013). ZnO nanoparticle-catalyzed efficient one-pot three-component synthesis of 3,4,5-trisubstituted furan-2(5H)-ones. *Journal of the Iranian Chemical Society*, *10(6)*, 1271–1277.

28. Goswami, P. (2009). Dually activated organo- and nano-cocatalyzed synthesis of coumarin derivatives. *Synthetic Communications*, *39(13)*, 2271–2278.

29. Kumar, B. V., Naik, H. S. B., Girija, D., & Kumar, B. V. (2011). ZnO nanoparticle as catalyst for efficient green one-pot synthesis of coumarins through Knoevenagel condensation. *Journal of Chemical Sciences*, *123(5)*, 615–621.

30. Ghomi, J. S., & Ghasemzadeh, M. A. (2014). A simple and efficient synthesis of 12-aryl-8,9,10,12-tetrahydrobenzo[a]xanthen-11-ones by ZnO nanoparticles catalyzed three component coupling reaction of aldehydes, 2-naphthol and dimedone. *South African Journal of Chemistry 67*, 27–32.

31. Tekale, S. U., Kauthale, S. S., Jadhav, K. M., & Pawar, R. P. (2013). Nano-ZnO catalyzed green and efficient one-pot four-component synthesis of pyranopyrazoles. *Journal of Chemistry*, *2013*, 1–8.

32. Siddiqui, Z. N., Ahmed, N., Farooq, F., & Khan, K. (2013). Highly efficient solvent-free synthesis of novel pyranyl pyridine derivatives via β-enaminones using ZnO nanoparticles. *Tetrahedron Letters*, *54(28)*, 3599–3604.

33. Haerizade, B. N., & Kassaee, M. Z. (2014). Nano ZnO promoted synthesis of 1,3-oxazoline-2-thione derivatives. *Journal of Chemical Research*, *38*, 295–296.

34. Krishnakumar, V., Mohan Kumar, K., Mandal, B. K., & Khan, F.-R. N. (2012). Zinc oxide nanoparticles catalyzed condensation reaction of isocoumarins and 1,7-heptadiamine in the formation of bis-isoquinolinones. *The Scientific World Journal*, *2012*, 1–7.

35. Baghbanian, S. M., Farhang, M., & Baharfar, R. (2011). One-pot three-component synthesis of α-amino nitriles catalyzed by nano powder TiO$_2$P$_{25}$. *Chinese Chemical Letters*, *22(5)*, 555–558.

36. Bharathi, A., Roopan, S. M., Kajbafvala, A., Padmaja, R. D., Darsana, M. S., & Nandhini Kumari, G. (2014). Catalytic activity of TiO$_2$ nanoparticles in the synthesis of some 2,3-disubstituted dihydroquinazolin-4(1H)-ones. *Chinese Chemical Letters*, *25(2)*, 324–326.

37. Kantam, M. L., Laha, S., Yadav, J., & Sreedhar, B. (2006). Friedel–Crafts alkylation of indoles with epoxides catalyzed by nanocrystalline titanium(IV) oxide. *Tetrahedron Letters*, *47(35)*, 6213–6216.

38. Sarkar, R., & Mukhopadhyay, C. (2013). Admicellar catalysis in multicomponent synthesis of polysubstituted pyrrolidinones. *Tetrahedron Letters*, *54*(*28*), 3706–3711.
39. Bahrami, K., Khodaei, M. M., & Naali, F. (2015). TiO$_2$ nanoparticles catalysed synthesis of 2-arylbenzimidazoles and 2-arylbenzothiazoles using hydrogen peroxide under ambient light. *Journal of Experimental Nanoscience*, *11*(*2*), 148–160.
40. Sachdeva, H., Dwivedi, D., Bhattacharjee, R. R., Khaturia, S., & Saroj, R. (2013). NiO nanoparticles: An efficient catalyst for the multicomponent one-pot synthesis of novel spiro and condensed indole derivatives. *Journal of Chemistry*, *2013*, 1–10.
41. Tanna, J., Chaudhary, R. G., Gandhare, N. V., Rai, A. R., & Juneja, H. D. (2015). Nickel oxide nanoparticles: Synthesis, characterization and recyclable catalyst, *International Journal of Scientific and Engineering Research*, 6, 93–98.
42. Nasseri, M. A., Ahrari, F., & Zakerinasab, B. (2015). Nickel oxide nanoparticles: A green and recyclable catalytic system for the synthesis of diindolyloxindole derivatives in aqueous medium. *RSC Advances*, *5*(*18*), 13901–13905.
43. Shiraishi, Y., Sugano, Y., Tanaka, S., & Hirai, T. (2010). One-pot synthesis of benzimidazoles by simultaneous photocatalytic and catalytic reactions on Pt@TiO$_2$ nanoparticles. *Angewandte Chemie International Edition*, *122*, 1700–1704.
44. Darehkordi, A., Ramezani, M., & Rahmani, F. (2018). TiO$_2$-nanoparticles catalyzed synthesis of new trifluoromethyl-4,5-dihydro-1,2,4-oxadiazoles and trifluoromethyl-1,2,4-oxadiazoles. *Journal of Heterocyclic Chemistry*, *55*(*7*), 1702–1708.
45. Rana, S., Brown, M., Dutta, A., Bhaumik, A., & Mukhopadhyay, C. (2013). Site-selective multicomponent synthesis of densely substituted 2-oxo dihydropyrroles catalyzed by clean, reusable, and heterogeneous TiO$_2$ nanopowder. *Tetrahedron Letters*, *54*(*11*), 1371–1379.
46. Khazaei, A., Reza Moosavi-Zare, A., Mohammadi, Z., Zare, A., Khakyzadeh, V., & Darvishi, G. (2013). Efficient preparation of 9-aryl-1,8-dioxo-octahydroxanthenes catalyzed by nano-TiO$_2$ with high recyclability. *RSC Advance*, *3*(*5*), 1323–1326.
47. Prasad, S. S., Jayaprakash, S. H., Rao, K. U., Reddy, N. B., Kumar, P. C. R., & Reddy, C. S., (2014). Nano-TiO$_2$ catalyzed microwave synthesis of α-hydroxyphosphonates. *Organic Communications*, *7*(*3*), 98–105.
48. Raghul, M. S., (2019). Efficient synthesis of RuO$_2$ nanoparticle with excellent activity for one-pot synthesis of 2,3-disubstituted quinazolin-4(1H)-ones. *Vietnam Journal of Chemistry*, *57*(*5*), 584–594.
49. Bento, A., Sanches, A., Medina, E., Nunes, C. D., & Vaz, P. D. (2015). MoO$_2$ nanoparticles as highly efficient olefin epoxidation catalysts. *Applied Catalysis A: General*, *504*, 399–407.
50. Li, X., Zhou, L., Gao, J., Miao, H., Zhang, H., & Xu, J. (2009). Synthesis of Mn$_3$O$_4$ nanoparticles and their catalytic applications in hydrocarbon oxidation. *Powder Technology*, *190*(*3*), 324–326.
51. Huang, C., Zhang, H., Sun, Z., Zhao, Y., Chen, S., Tao, R., & Liu, Z. (2011). Porous Fe$_3$O$_4$ nanoparticles: Synthesis and application in catalyzing epoxidation of styrene. *Journal of Colloid and Interface Science*, *364*(*2*), 298–303.
52. Suresh, L., Vijay Kumar, P. S., Vinodkumar, T., & Chandramouli, G. V. P. (2016). Heterogeneous recyclable nano-CeO$_2$ catalyst: Efficient and eco-friendly synthesis of novel fused triazolo and tetrazolo pyrimidine derivatives in aqueous medium. *RSC Advances*, *6*(*73*), 68788–68797.
53. Maleki, A., & Firouzi-Haji, R. (2018). L-Proline functionalized magnetic nanoparticles: A novel magnetically reusable nanocatalyst for one-pot synthesis of 2,4,6-triarylpyridines. *Scientific Reports*, *8*(*1*), 1–8.

54. Payra, S., Saha, A., & Banerjee, S. (2018). Magnetically recoverable $Fe_3 O_4$ nanoparticles for the one-pot synthesis of coumarin-3-carboxamide derivatives in aqueous ethanol. *ChemistrySelect, 3(26)*, 7535–7540.

55. Saha, A., Payra, S., Selvaratnam, B., Bhattacharya, S., Pal, S., Koodali, R. T., & Banerjee, S. (2018). Hierarchical mesoporous RuO_2/Cu_2O nanoparticle-catalyzed oxidative homo/hetero azo-coupling of anilines. *ACS Sustainable Chemistry & Engineering, 6*, 11345–11352.

56. Tanna, J. A., Chaudhary, R. G., Sonkusare, V. N., & Juneja, H. D. (2016). CuO nanoparticles: Synthesis, characterization and reusable catalyst for polyhydroquinoline derivatives under ultrasonication. *Journal of the Chinese Advanced Materials Society, 4(2)*, 110–122.

57. Girija, D., Halehatty, S., Bhojya. S. N., Sudhamani, C. N., & Vinay, K. B. (2011). Cerium oxide nanoparticles – a green, reusable, and highly efficient heterogeneous catalyst for the synthesis of Polyhydroquinolines under solvent-free conditions, *Archives of Applied Science Research, 3(3)*, 373–382.

58. Shahram, B. (2011). Efficient and reusable heterogeneous nano sized CeO_2 catalyst for the synthesis of 1,8-dioxo-octahydroxanthenes. *International Journal of Chemical and Biochemical Sciences, 6*, 72–75.

59. Ali, G., Javad, S. G., & Safura, Z. (2013). Fe_3O_4 nanoparticles: A highly efficient and easily reusable catalyst for the one-pot synthesis of xanthene derivatives under solvent-free conditions. *Journal of the Serbian Chemical Society, 78(6)*, 769–779.

60. Masoud, N. E., Jafar, H., & Fatemeh, M. (2011). Fe_3O_4 nanoparticles as an efficient and magnetically recoverable catalyst for the synthesis of 3,4-dihydropyrimidin-2(1H)-ones under solvent-free conditions. *Chinese Journal of Catalysis, 32*, 1484–1489.

61. Sajadi, S. M. (2013). Nano cerium oxide as a recyclable catalyst for the synthesis of N-monosubstituted ureas with the aid of acetaldoxime as an effective water surrogate. *Journal of Chemical Research, 37*, 623–625.

62. Ghomi, J. S., Ghasemzadeh, M. A., & Safura, Z. (2013). ZnO nanoparticles: A highly effective and readily recyclable catalyst for the one-pot synthesis of 1,8-dioxo-decahydroacridine and 1,8-dioxooctahydro-xanthene derivatives. *Journal of the Mexican Chemical Society, 57 (1)*, 1–7.

63. Dighore, N. R., Anandgaonker, P. L., Gaikwad, S. T., & Rajbhoj, A. S. (2014). Solvent free green synthesis of 5-arylidine barbituric acid derivatives catalyzed by copper oxide nanoparticles. *Research Journal of Chemical Sciences, 4(7)*, 93–98.

64. Manju, K., Smitha, T., Divya, S. N., Aswathy, E. K., Aswathy, B., Arathy, T., & Krishna, K. T. (2015). Structural, magnetic, and acidic properties of cobalt ferrite nanoparticles synthesised by wet chemical methods. *Journal of Advance Ceramics, 4*, 199–205.

65. Swatsitang, E., Phokha, S., Hunpratub, S., Usher, B., Bootchanont, A., Maensiri, S., & Chindaprasirt, P. (2016). Characterization and magnetic properties of cobalt ferrite nanoparticles. *Journal of Alloys and Compound, 664*, 792–797.

66. Yang, Y., Jing, L., Yu, X., Yan, D., & Gao, M. (2007). Coating aqueous quantum dots with silica via reverse microemulsion method: Toward size-controllable and robust fluorescent nanoparticles. *Chemistry of Materials, 19*, 4123–4128.

67. Sanpo, N., Berndt, C. C., & Wang, J. (2012). Microstructural and antibacterial properties of zinc-substituted cobalt ferrite nanopowders synthesized by sol-gel methods. *Journal of Applied Physics, 112*, 084333.

68. Sattar, A. A., Sayed, H. M., & Ibrahim, A. L. S. (2015). Structural and magnetic properties of $CoFe_2O_4/NiFe_2O_4$ core/shell nanocomposite prepared by the hydrothermal method. *Journal of Magnetism and Magnetic Materials, 395*, 89–96.

69. Vilar, Y. S., Andujar, S. M., Aguirre, G. C., Mira, J., Rodriguez, M. A., & Garcia, C. S. (2009). A simple solvothermal synthesis of MFe_2O_4 (M = Mn, Co and Ni) nanoparticles. *Journal of Solid State Chemistry, 182*, 2685–2690.
70. Molazemi, M., Shokrollahi, H., & Hashemi, B. (2013). The investigation of the compression and tension behavior of the cobalt ferrite magnetorheological fluids synthesized by co-precipitation. *Journal of Magnetism and Magnetic Material, 346*, 107–112.
71. Hu, X. G., & Dong, S. J. (2008). Metal nanomaterials and carbon nanotubes synthesis, functionalization and potential applications towards electro-chemistry. *Journal of Material Chemistry, 18*, 1279–1295.
72. Chagas, C. A., de Souza, E. F., de Carvalho, M. C. N. A., Martins, R. L., & Schmal, M. (2016). Cobalt ferrite nanoparticles for the preferential oxidation of CO. *Applied Catalysis A: General, 519*, 139–145.
73. Sadri, F., Ramazani, A., Massoudi, A., Khoobi, M., Azizkhani, V., Tarasi, R., Dolatyari, L., & Min, B. K. (2014). Magnetic $CoFe_2O_4$ nanoparticles as an efficient catalyst for the oxidation of alcohols to carbonyl compounds in the presence of oxone as an oxidant. *Bull Korean Chemistry Society, 35*, 2029–2032.
74. Tong, J., Bo, L., Li, Z., Lei, Z., & Xia, C. (2009). Magnetic $CoFe_2O_4$ nanocrystal: A novel and efficient heterogeneous catalyst for aerobic oxidation of cyclohexane. *Journal of Molecular Catalysis A: Chemical, 307(1–2)*, 58–63.
75. Moghaddam, F. M., Tavakoli, G., & Aliabadi, A. (2015). Application of nickel ferrite and cobalt ferrite magnetic nanoparticles in C–O bond formation: A comparative study between their catalytic activities. *RSC Advances, 5(73)*, 59142–59153.
76. Kazemi, M., Ghobadi, M., & Mirzaie, A. (2018). Cobalt ferrite nanoparticles ($CoFe_2O_4$ MNPs) as catalyst and support: Magnetically recoverable nanocatalysts in organic synthesis. *Nanotechnology Reviews, 7(1)*, 43–68.
77. Liu, Y., Zhou, L., Hui, X., Dong, Z., Zhu, H., Shao, Y., & Li, Y. (2014). Fabrication of magnetic amino-functionalized nanoparticles for S-arylation of heterocyclic thiols. *RSC Advances, 4(90)*, 48980–48985.
78. Li, P. H., Li, B. L., An, Z. M., Mo, L. P., Cui, Z. S., & Zhang, Z. H. (2013). Magnetic nanoparticles ($CoFe_2O_4$)-supported phosphomolybdate as an efficient, green, recyclable catalyst for synthesis of β-hydroxy hydroperoxides. *Advanced Synthesis and Catalysis, 355*, 2952–2959.
79. Senapati, K. K., Borgohain, C., & Phukan, P. (2011). Synthesis of highly stable $CoFe_2O_4$ nanoparticles and their use as magnetically separable catalyst for Knoevenagel reaction in aqueous medium. *Journal of Molecular Catalysis A: Chemical, 339*, 24–31.
80. Content, S., Dupont, T., Fédou, N. M., Penchev, R., Smith, J. D., Susanne, F., Stoneley, C., Twiddle, S. J. R. (2013). Optimization of the manufacturing route to PF-610355 (2): Synthesis of the API. *Organic Process Research & Development, 17(2)*, 202–212.
81. Zhao, X. N., Hu, H. C., Zhang, F. J., & Zhang, Z. H. (2014). Magnetic $CoFe_2O_4$ nanoparticle immobilized N-propyl diethylenetriamine sulfamic acid as an efficient and recyclable catalyst for the synthesis of amides via the Ritter reaction. *Applied Catalysis A: General, 482*, 258–265.
82. Shaikh, M., Sahu, M., Atyam, K. K., & Ranganath, K. V. S. (2016). Surface modification of ferrite nanoparticles with dicarboxylic acids for the synthesis of 5-hydroxymethylfurfural: A novel and green protocol. *RSC Advances, 6(80)*, 76795–76801.
83. Masunga, N., Kelebogile Mmelesi, O., Kefeni, K. K., & Mamba, B. B. (2019). Recent advances in copper ferrite nanoparticles and nanocomposites synthesis, magnetic properties and application in water treatment: Review. *Journal of Environmental Chemical Engineering, 7*, 103179.

84. Viswanath, I. V. K., & Murthy, Y. L. N. (2013). One-pot, three-component synthesis of 1,4-dihydropyridines by using nano crystalline copper ferrite. *Chemical Science Transaction*, 2 (*1*), 227–233.

85. Gharib, A., Noroozi Pesyan, N., Vojdani Fard, L., & Roshani, M. (2014). Catalytic synthesis of α-aminonitriles using nano copper ferrite $CuFe_2O_4$ under green conditions. *Organic Chemistry International*, *2014*, 1–8.

86. Yang, S., Wu, C., Zhou, H., Yang, Y., Zhao, Y., Wang, C., Yang, W., Xu, J. (2012). An Ullmann C–O coupling reaction catalyzed by magnetic copper ferrite nanoparticles. *Advanced Synthesis & Catalysis*, *355(1)*, 53–58.

87. Murthy, Y. L. N., Diwakar, B. S., Govindh, B., Nagalakshmi, K., Viswanath, I. V. K., & Singh, R. (2012). Nano copper ferrite: A reusable catalyst for the synthesis of β,γ-unsaturated ketones. *Journal of Chemical Sciences*, *124(3)*, 639–645.

88. Satish, G., Reddy, K. H. V., Anil, B. S. P., Shankar, J., Uday Kumar, R., & Nageswar, Y. V. D. (2014). Direct C–H amination of benzothiazoles by magnetically recyclable $CuFe_2O_4$ nanoparticles under ligand-free conditions. *Tetrahedron Letters*, *55(40)*, 5533–5538.

89. Murthy, Y. L. N., Mahesh, P., Devi, B. R.; Durgeswarai, L. K., & Mani, P. (2014). Synthesis and antibacterial assay of 9-substituted aryl-1,8-dioxo-octahydroxanthenes. *Asian Journal of Chemistry*, *26(15)* 4594–4598.

90. Nguyen, A. T., Pham, L. T., Phan, N. T. S., & Truong, T. (2014). Efficient and robust superparamagnetic copper ferrite nanoparticle-catalyzed sequential methylation and C–H activation: Aldehyde-free propargylamine synthesis. *Catalysis Science and Technology*, *4(12)*, 4281–4288.

91. Srinivas, B. T. V., Rawat, V. S., Konda, K., & Sreedhar, B. (2014). Magnetically separable copper ferrite nanoparticles-catalyzed synthesis of diaryl, alkyl/aryl sulfones from arylsulfinic acid salts and organohalides/boronic acids. *Advanced Synthesis & Catalysis*, *356(4)*, 805–817.

92. Panda, N., Jena, A. K., Mohapatra, S., & Rout, S. R. (2011). Copper ferrite nanoparticle-mediated N-arylation of heterocycles: A ligand-free reaction. *Tetrahedron Letters*, *52(16)*, 1924–1927.

93. Sarode, S. A., Bhojane, J. M., & Nagarkar, J. M. (2015). An efficient magnetic copper ferrite nanoparticle: For one pot synthesis of 2-substituted benzoxazole via redox reactions. *Tetrahedron Letters*, *56(1)*, 206–210.

94. Nguyen, A. T., Nguyen, L. T. M., Nguyen, C. K., Truong, T., & Phan, N. T. S. (2014). Superparamagnetic copper ferrite nanoparticles as an efficient heterogeneous catalyst for the α-arylation of 1,3-diketones with C-C cleavage. *ChemCatChem*, *6(3)*, 815–823.

95. Gholinejad, M., & Aminianfar, A. (2015). Palladium nanoparticles supported on magnetic copper ferrite nanoparticles: The synergistic effect of palladium and copper for cyanation of aryl halides with $K_4[Fe(CN)_6]$. *Journal of Molecular Catalysis A: Chemical*, *397*, 106–113.

96. Khalili, D., Rezaei, M., & Koohgard, M. (2019). Ligand-free copper-catalyzed O-arylation of aryl halides using impregnated copper ferrite on mesoporous graphitic carbon nitride as a robust and magnetic heterogeneous catalyst. *Microporous and Mesoporous Materials*, *287*, 254–263.

97. Gholinejad, M., & Ahmadi, J. (2015). Assemblies of copper ferrite and palladium nanoparticles on silica microparticles as a magnetically recoverable catalyst for Sonogashira reaction under mild conditions. *ChemPlusChem*, *80(6)*, 973–979.

98. Gholinejad, M., Karimi, B., & Mansouri, F. (2014). Synthesis and characterization of magnetic copper ferrite nanoparticles and their catalytic performance in one-pot odorless carbon-sulfur bond formation reactions. *Journal of Molecular Catalysis A: Chemical*, *386*, 20–27.

99. Dixit, G., Singh, J. P., Srivastava, R. C., Agrawal, H. M., & Gupta, A. (2010). Structural and magnetic behavior of $NiFe_2O_4$ thin film grown by pulse laser deposition. *Indian Journal of Pure and Applied Physics*, *48*, 287–291.
100. Borade, R. M., Somvanshi, S. B., Kale, S. B., Pawar, R. P., & Jadhav, K. M. (2020). Spinel zinc ferrite nanoparticles: An active nanocatalyst for microwave irradiated solvent free synthesis of chalcones. *Materials Research Express*, *7*, 1–32.
101. Jadhav, S. V., Jinka, K. M., & Bajaj, H. C. (2012). Nanosized sulfated zinc ferrite as catalyst for the synthesis of nopol and other fine chemicals. *Catalysis Today*, *198(1)*, 98–105.
102. Matloubi Moghaddam, F., Tavakoli, G., & Rezvani, H. R. (2014). Highly active recyclable heterogeneous nanonickel ferrite catalyst for cyanation of aryl and heteroaryl halides. *Applied Organometallic Chemistry*, *28(10)*, 750–755.
103. Kulkarni, A. M., Desai, U. V., Pandit, K. S., Kulkarni, M. A., & Wadgaonkar, P. P. (2014). Nickel ferrite nanoparticles–hydrogen peroxide: A green catalyst-oxidant combination in chemoselective oxidation of thiols to disulfides and sulfides to sulfoxides. *RSC Advances*, *4(69)*, 36702.
104. Ghahremanzadeh, R., Rashid, Z., Zarnani, A. H., & Naeimi, H. (2014). Manganese ferrite nanoparticle catalyzed tandem and green synthesis of spirooxindoles. *RSC Advances*, *4(82)*, 43661–43670.
105. Alhadlaq, H. A., Akhtar, M. J., & Ahamed, M. (2015). Zinc ferrite nanoparticle-induced cytotoxicity and oxidative stress in different human cells. *Cell & Bioscience*, *5(1)*, 1–11.
106. Misra, R. D. K., Kale, A., Srivastava, R. S., & Senkov, O. N. (2003). Synthesis of nanocrystalline nickel and zinc ferrites by microemulsion technique. *Materials Science and Technology*, *19(6)*, 826–830.
107. Shenoy, S. D., Joy, P. A., & Anantharaman, M. R. (2004). Effect of mechanical milling on the structural, magnetic and dielectric properties of coprecipitated ultrafine zinc ferrite. *Journal of Magnetism and Magnetic Materials*, *269(2)*, 217–226.
108. Papa, F., Patron, L., Carp, O., Paraschiv, C., & Ioan, B. (2009). Catalytic activity of neodymium substituted zinc ferrites for oxidative conversion of methane. *Journal of Molecular Catalysis A: Chemical*, *299(1–2)*, 93–97.
109. Amigó, R., Asenjo, J., Krotenko, E., Torres, F., Tejada, J., & Brillas, E. (2000). electrochemical synthesis of new magnetic mixed oxides of Sr and Fe: Composition, magnetic properties, and microstructure. *Chemistry of Materials*, *12(2)*, 573–579.
110. Rodriguez, J. A., Hanson, J. C., Chaturvedi, S., Maiti, A., & Brito, J. L. (2000). Phase transformations and electronic properties in mixed metal oxides; experimental and theoretical studies on the behavior of $NiMoO_4$ and $MgMoO_4$. *Journal of Chemical Physics*, *112 (2)*, 935–945.
111. (a) Cui, H., Zayat, M., & Levy, D. (2005). Sol-gel synthesis of nanoscaled spinels using propylene oxide as a gelation agent. *Journal of Sol-Gel Science and Technology*, *35(3)*, 175–181; (b) Elia, A., Martin Aispuro, P., Quaranta, N., MartınMartınez, J. M., & Vazquez, P. (2011). *Macromolecular Symposia*, *301*, 136; (c) Kim, Y. J., Rawal, S. B., Sung, S. D., & Lee, W. I. (2011). *Bulletin of the Korean Chemical Society*, *32*, 141.
112. (a) Reddy, B. M., Chowdhury, B., & Smirniotis, P. G. (2001). An XPS study of the dispersion of MoO_3 on TiO_2–ZrO_2, TiO_2–SiO_2, TiO_2–Al_2O_3, SiO_2–ZrO_2, and SiO_2–TiO_2–ZrO_2 mixed oxides. *Applied Catalysis A: General*, *211(1)*, 19–30; (b) Sankar, G., Rao, C. N. R., & Rayment, T., *Journal of Materials Chemistry*, 1991, *1*, 299.
113. (a) Tang, A., Yang, H., & Zhang, X. (2006). *International Journal of Physical Sciences*, *1*, 101; (b) Zyryanov, V. V. (2003). *Inorganic Materials*, *39*, 1163.
114. (a) Ajaikumar, S., & Pandurangan, A. (2009). Efficient synthesis of quinoxaline derivatives over ZrO_2/M_xO_y (M = Al, Ga, In and La) mixed metal oxides supported on MCM-41

mesoporous molecular sieves. *Applied Catalysis A: General, 357(2)*, 184–192; (b) Sheets, W. C., Stampler, E. S., Kabbour, H., Bertoni, M. I., Cario, L., Mason, T. O., Marks, T. B., & Poeppelmeier, K. R. (2007). facile synthesis of BiCuOS by hydrothermal methods. *Inorganic Chemistry, 46(25)*, 10741–10748.

115. (a) Jiang, D., Su, L., Ma, L., Yao, N., Xu, X., Tang, H., & Li, X. (2010). *Applied Surface Science, 256*, 3216; (b) Reddy, B. M., & Ganesh, I. (2001). *Journal of Molecular Catalysis A: Chemical, 169*, 207.

116. Biradar, A. V., Umbarkar, S. B., & Dongare, M. K. (2005). Transesterification of diethyl oxalate with phenol using MoO_3/SiO_2 catalyst. *Applied Catalysis A: General, 285(1–2)*, 190–195.

117. Singh, S. J., & Jayaram, R. V. (2008). Chemoselective O-tert-butoxycarbonylation of hydroxy compounds using $NaLaTiO_4$ as a heterogeneous and reusable catalyst. *Tetrahedron Letters, 49(27)*, 4249–4251.

118. Gawande, M. B., & Jayaram, R. V. (2006). *Catalysis Communications, 7*, 931.

119. Khaleel, A., & Richards, R. M. (2009). In *Nanoscale Materials in Chemistry*. K. J. Klabunde (Ed.), (pp. 85–120). Wiley Interscience: New York.

120. (a) Smith, M. B., & March, J. (2001). *March's Advanced Organic Chemistry: Reactions, Mechanisms and Structure*, John Wiley & Sons: New York; (b) Norcross, R. D., & Paterson, I. (1995). Total synthesis of bioactive marine macrolides. *Chemical Reviews, 95(6)*, 2041–2114; (c) Trost, B. M., & Fleming, I. (1991). *Comprehensive Organic Synthesis: 9-Volume Set*, Pergamon: Oxford, Vol. 2, Parts 1.4–1.7.

121. Gawande, M. B., Branco, P. S., Parghi, K., Shrikhande, J. J., Pandey, R. K., Ghumman, C. A. A., Teodoro, O. M. N. D., & Jayaram, R. V. (2011). Synthesis and characterization of versatile $MgO–ZrO_2$ mixed metal oxide nanoparticles and their applications. *Catalysis Science & Technology, 1(9)*, 1653–1664.

122. Papageorgiou, E. A., Gaunt, M. J., Yu, J., & Spencer, J. B. (2000). Selective hydrogenolysis of novel benzyl carbamate protecting groups. *Organic Letters, 2(8)*, 1049–1051.

123. Chandorkar, J. G., Umbarkar, S. B., Rode, C. V., Kotwal, V. B., & Dongare, M. K. (2007). Synthesis of tinidazole by condensation–oxidation sequence using MoO_3/SiO_2 bifunctional catalyst. *Catalysis Communications, 8(10)*, 1550–1555.

124. Sultana, S. S. P., Kishore, D. H. V., Kuniyil, M., Khan, M., Alwarthan, A., Prasad, K. R. S., Labis, J. P., & Adil, S. F. (2015). Ceria doped mixed metal oxide nanoparticles as oxidation catalysts: Synthesis and their characterization. *Arabian Journal of Chemistry, 8(6)*, 766–770.

125. Palacio, M., Villabrille, P., Romanelli, G., Vázquez, P., & Cáceres, C. (2010). Ecofriendly catalysts based on mixed xerogels for liquid phase oxidations by hydrogen peroxide. *Studies in Surface Science and Catalysis*, 425–428. Elsevier: Amsterdam

126. Sapkal, B. M., Labhane, P., Disale, S. T., & More, D. H. (2019). $ZnO@SnO_2$ mixed metal oxide as an efficient and recoverable nanocatalyst for the solvent free synthesis of hantzsch 1,4-dihydropyridines. *Letters in Organic Chemistry, 16(2)*, 139–144.

3 Noble Metal Nanoparticles in Organic Catalysis

Laxmi Devi, Komal, Sunita Kanwar, Kamal Nayan Sharma, and Anirban Das
Amity University Haryana, Gurugram (Haryana), India

Jyotirmoy Maity
University of Delhi, Delhi, India

CONTENTS

DOI: 10.1201/9781003126270-3

3.1   INTRODUCTION

In the last decades of the 20th-century nanoscience has emerged as a versatile field with diverse applications [1]. Nanomaterials have been used as efficient and selective catalysts for many organic reactions due to their unique properties like high surface area, specific surface termination (specific atoms on surface) and quantum confinement [2]. Tailoring the size and morphology of these nanocatalysts resulted in further tuning of the efficiency and selectivity of these catalysts [3]. Thus, synthetic methodologies played an important role in the development of the overall catalytic process. A variety of physical and chemical techniques were utilized to investigate the synthesis of these metal nanoparticles (NPs) [4]. Apart from targeting NPs with more favourable properties, much effort was also being directed nowadays making the syntheses processes environmentally benevolent [5]. Noble metal NPs have the advantage of being chemically unreactive towards various chemical reagents, thus remain stable during the course of the reactions and may be used for a large number of catalytic cycles without loss of activity.

Au NPs were very important and unique in nature because of their physicochemical properties that can be tuned by proper control of shape and size [6]. Applications of the Au NPs towards the total synthesis of natural products have been reviewed [7]. Pd NPs were used extensively as catalysts due to their excellent physicochemical properties such as high thermal stability and good chemical stability [8]. Ultrasmall Ir NPs were very efficient as well as selective for reactions like the catalytic reduction of nitroarenes under mild conditions [9]. The novel Ir-based nanomaterials were reported as good heterogeneous catalysts for industrial applications [10]. Pt had high corrosion resistance and numerous catalytic applications including in automotive catalytic converters and petrochemical cracking [11]. Ag was found to be an extremely useful noble metal due to their useful characteristics such as antimicrobial activity in addition to their suitability as industrial heterogeneous for many organic transformations [12]. This book chapter discusses the application of noble metal NPs in organic catalytic transformations.

3.2 ORGANIC CATALYSIS BY USING AU NP

Au NPs were used widely as catalysts in organic transformations. They were utilized in many reactions such as nucleophilic addition reactions, oxidation reactions and hydrogenation reactions.

3.2.1 OXIDATION REACTIONS

Oxidation reactions catalysed by Au NPs were being widely studied and showed high selectivity and reactivity towards oxidation of molecules containing the alcohol functional group, which includes diols, polyols, and amino alcohols.

3.2.1.1 Oxidation of Aliphatic Alcohols

The catalyst Au supported on C (Au/C) was found to be selective towards oxidation of the aliphatic alcohol geraniol. The major products were *cis* and *trans*-citral although β-pinene, limonene, nerol, etc. along with trace quantities of geranic acid were observed (Figure 3.1). Au NPs (2–5 nm-sized) supported on nanocrystalline

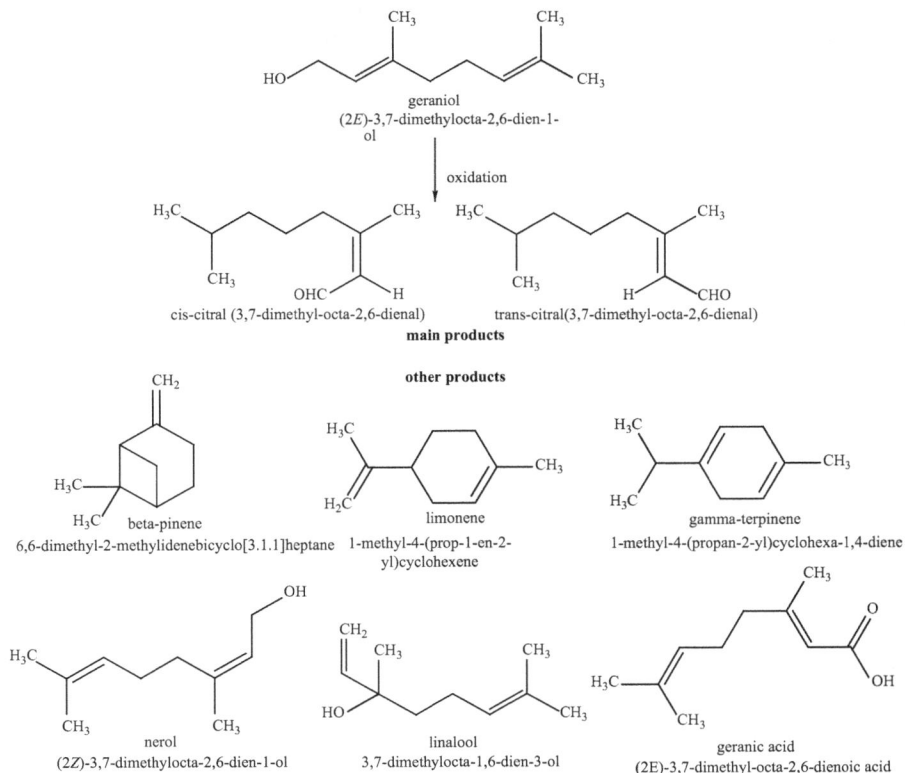

FIGURE 3.1 Oxidation reaction of geraniol (main products and some by-products) catalysed by Au/C.

ceria were found to be selective and recyclable catalysts for oxidizing allylic alcohol to form unsaturated ketones even in absence of base and solvent. [13]

3.2.1.2   Oxidation of Diols

Oxidation of ethane-1,2-diols to glycolic acid and propane-1,2-diols to lactic acid was carried out using Au NPs with high selectivity (Figure 3.2). Au/C was very active in the selective oxidation of 1,2-diols into α-hydroxy acids, under mild conditions with its activity being superior to mono-, di- and tri-metallic Pd-, Pt- and Bi-based catalysts, as Au afforded higher selectivity and enhanced resistance to poisoning. [14]

3.2.1.3   Oxidation of Polyols

Gluconic acid and gulonic acid were produced on oxidizing D-sorbitol using gold-platinum group metal on carbon support (Au-PGM/C) as a catalyst (Figure 3.3) [15]. It was seen that bimetallic catalysts exhibited improved selectivity as well as improved resistance to poisoning in comparison to monometallic catalysts. Selectivity observed for gluconate and gulonate on Au and Au/Pt were 60% and 62%, respectively [15, 16]. For the oxidation of glycerol, Au was found to be resistant to the poisoning of oxygen than PGMs. The oxidation of glycerol in the liquid phase using Au/charcoal, activated carbon, and Au/graphite was reported. It was found that glyceric acid was obtained by oxidation of glycerol with high selectivity and quantitative conversion in mild conditions using water as a solvent (Figure 3.4) [17].

3.2.1.4   Oxidation of Amino Alcohols

Prati and coworkers suggested that amino alcohols could be converted into amino acids in alkaline conditions by using Au NP [18]. Serinol and ethanolamine were also converted into polyols by using Au on different types of metal oxides such as Al_2O_3, TiO_2, $MgAl_2O_4$ and MgO [19].

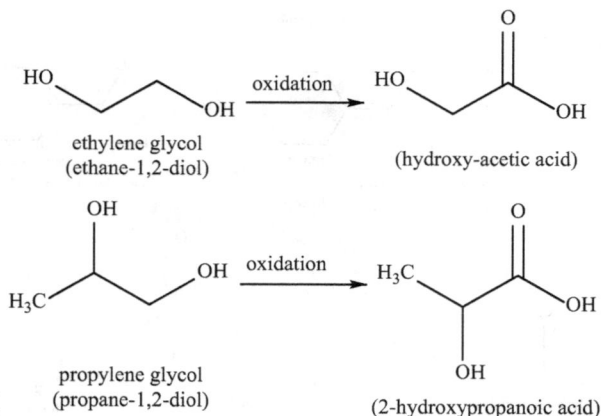

FIGURE 3.2   Selective oxidation reactions of diols using Au/C as catalyst.

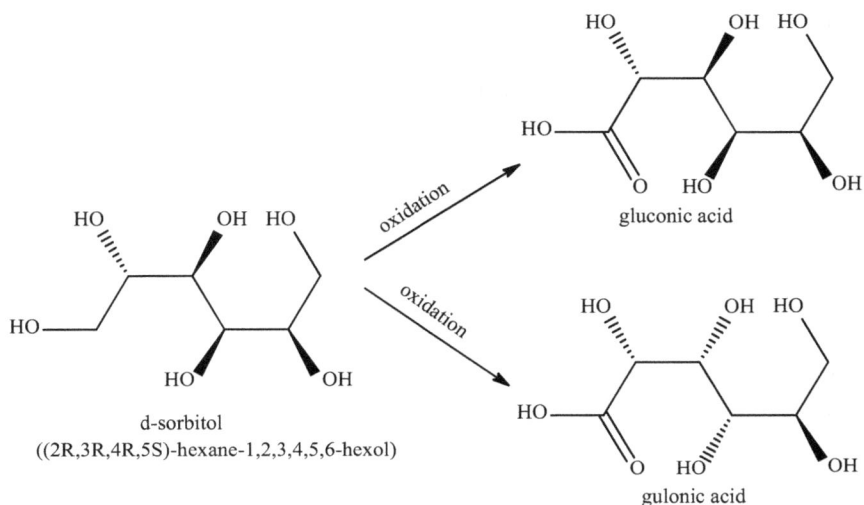

FIGURE 3.3 Oxidation reaction of D-sorbitol using Au-PGM/C as catalyst.

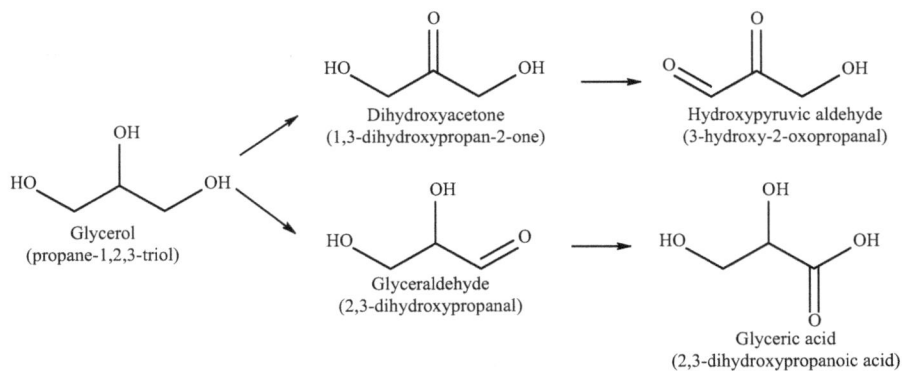

FIGURE 3.4 Oxidation reactions of glycerol catalysed by Au/charcoal.

3.2.1.5 Oxidation of Cycloalcohols

Cycloalcohols (e.g. cyclohexanol, cyclooctanol, cyclododecanol, 4-methyl cyclo-hexanol) were selectively oxidized to corresponding aldehydes/ketones using molecular oxygen (O_2) on Au NP deposited on SiO_2, TiO_2, Al_2O_3, Cr_2O_3, Fe_2O_3, CoO_x, MnO_2, CuO and NiO [20]. The nature of support and the synthesis proce-dure of the catalyst had a significant influence on the catalytic activity. The Au/CuO catalyst prepared using the co-precipitation method was found to be very selective at pH 10 and the reaction could be carried out *via* oxidative dehydroxylation by β-C–H elimination [14].

3.2.1.6 Oxidation of Aromatic Alcohols

Au NP (2–5 nm-sized) on ceria were used to oxidize alcohols like 2-hydroxybenzyl alcohol, 2-phenoxyethanol, vanillin alcohol, cinnamyl alcohol, 3-phenyl-1-propanol and 3,4-dimethyoxybenzyl alcohol (Figure 3.5) using molecular oxygen at atmospheric pressure without the use of a base or solvent [21–23]. Au–Pd NP dispersed on TiO$_2$/graphene oxide composites were used for oxidation of benzyl alcohol and 4-methoxybenzyl alcohol (Figures 3.6 and 3.7). Au NP on different metal oxides

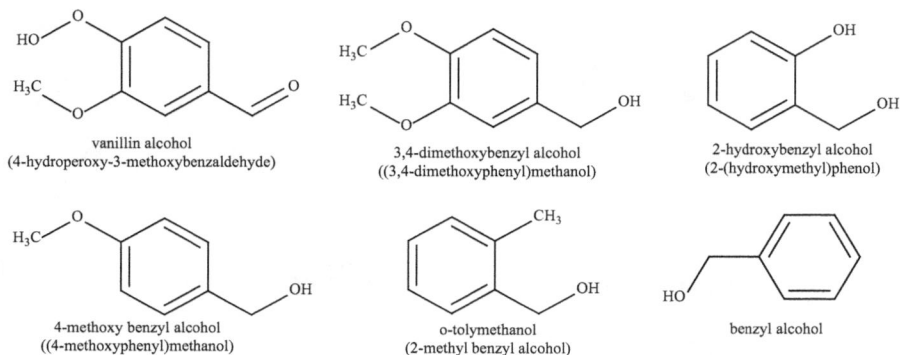

vanillin alcohol
(4-hydroperoxy-3-methoxybenzaldehyde)

3,4-dimethoxybenzyl alcohol
((3,4-dimethoxyphenyl)methanol)

2-hydroxybenzyl alcohol
(2-(hydroxymethyl)phenol)

4-methoxy benzyl alcohol
((4-methoxyphenyl)methanol)

o-tolymethanol
(2-methyl benzyl alcohol)

benzyl alcohol

FIGURE 3.5 Examples of different types of alcohols used in selective oxidation of alcohols using Au NP-based catalysts.

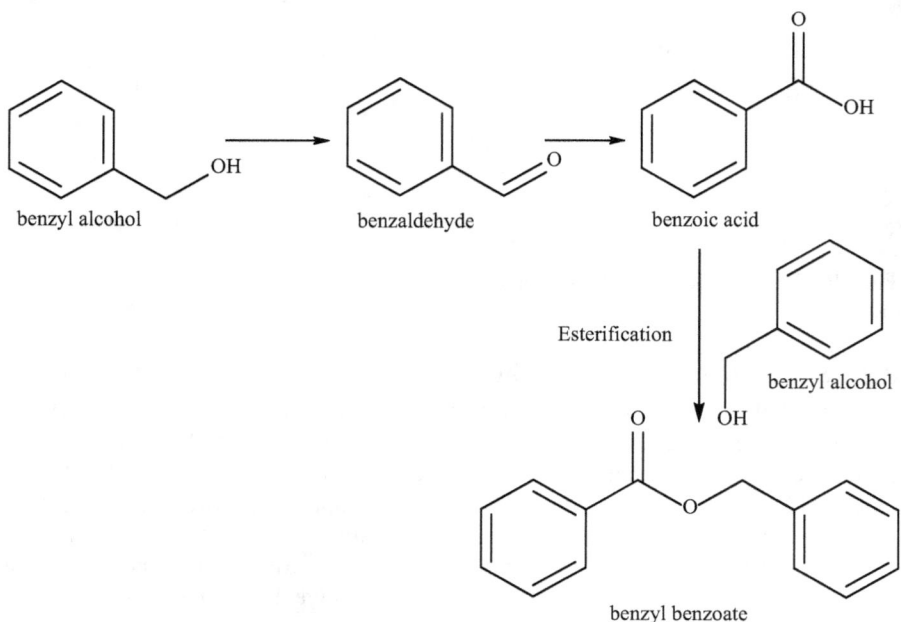

benzyl alcohol

benzaldehyde

benzoic acid

Esterification

benzyl alcohol

benzyl benzoate

FIGURE 3.6 Oxidation reaction of benzyl alcohol catalysed by Au/TiO$_2$ catalyst.

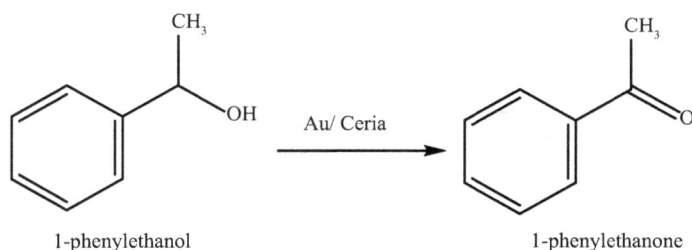

FIGURE 3.7 Partial oxidation reaction of methylbenzyl alcohol, a secondary alcohol to ketone using Au/Ceria as catalyst.

(TiO$_2$, ZnO, Fe$_2$O$_3$, MgO$_2$ and Al$_2$O$_3$,) was used to obtain benzaldehyde (from benzyl alcohol) and acetophenone (from methylbenzyl alcohol) using *tert*-butylhydroperoxide as an oxidant under microwave irradiation for 1 h, with no by-products [24]. During the oxidation of benzyl alcohol, some by-products like benzoic acid and benzyl benzoate were also formed. Benzoic acid was formed due to the over oxidation of benzaldehyde and benzyl benzoate was formed because of additional esterification in the presence of benzoic acid and benzyl alcohol [14, 25].

3.2.1.7 Oxidation of Alkanes

3.2.1.7.1 Oxidation of Cyclohexane to Adipic Acid

Adipic acid was synthesized at an industrial level in two steps starting from cyclohexane. In the first step, the cyclohexane was converted into cyclohexanol and cyclohexanone (KA-oil). This step was carried out in the presence of O$_2$ under 10–20 bar pressure using Co- and Mn-based catalysts. In the second step, KA-oil was treated with nitric acid which formed adipic acid (Figure 3.8). This method had some drawbacks like having a low conversion (4%–12%) and use of ecologically toxic reagents. Catalysts like Au/TiO$_2$ showed only 16.4%–21.6% selectivity. Alshammari *et al.* compared Au supported on various oxide supports and described the selectivity order and catalytic activity. The relative order of activity was found to be TiO$_2$ > Al$_2$O$_3$ > ZrO$_2$ > MgO > CaO. This reaction was favoured by higher temperature but temperature above 423 K produced more amounts of undesirable products than desired products [26].

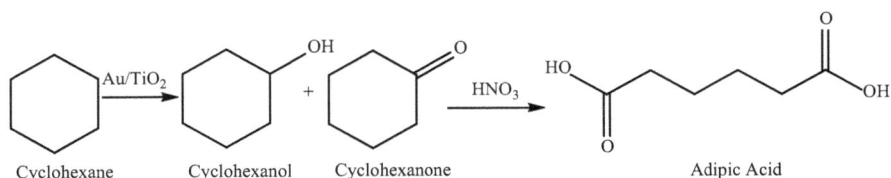

FIGURE 3.8 Oxidation of cyclohexane catalysed using Au/TiO$_2$.

3.2.2   Epoxidation Reaction

Epoxidation of alkenes like ethene, cyclohexene, styrene, etc. was relatively facile whereas epoxidation of propene was challenging due to the presence of allylic hydrogen. The first report of propene being converted into propylene oxide was by Haruta *et al.* who used Au NP in the presence of O_2 and sacrificial H_2. [7] However, a recent report showed that sacrificial H_2 was not necessary for epoxidation (Figure 3.9). [27]

3.2.3   Hydrogenation Reactions

3.2.3.1   Hydrogenation Reactions of Alkenes, Dienes and Alkynes

Erkelens *et al.* [28] used an Au film to catalyse the hydrogenation of cyclohexene at 469–615 K. Later, the hydrogenation reaction for 1,3-butadiene and 2-butyne using Au on SiO_2 was demonstrated by Bond *et al.* in 1973 (Figure 3.10) [29]. In the monohydrogenation reaction of 1,3-butadiene the chemoselectivity was observed without any diastereoselectivity and in the hydrogenation reaction of 2-butyne, diastereoselectivity (8:1) was observed [7].

3.2.3.2   Hydrogenation of α,β-unsaturated Carbonyl Groups

Supported Au NPs showed high selectivity towards hydrogenation of α,β-unsaturated carbonyl groups. The formation of crotyl alcohol (R_2 = H, R_1 and R_3 = CH_3) from crotonaldehyde proceeded with 81% selectivity and 5%–10% conversion was achieved by using Au/ZnO_2 or Au/ZnO catalysts (Figure 3.11) [30]. The Au catalyst showed high selectivity towards C=O over C=C. Hydrogenation of acrolein (R_1, R_2, R_3 = H) was also reported using AuO_2, Au/ZrO_2, Au/TiO_2 and Au-In/ZnO [31] A summary of reported reactions that Au-NPs catalyse is given in Table 3.1.

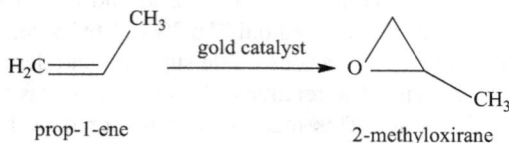

FIGURE 3.9   Epoxidation of propene catalysed by Au.

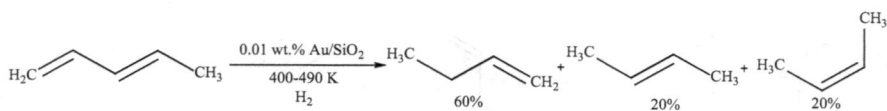

FIGURE 3.10   Hydrogenation of 1,3-butadiene catalysed by Au/SiO_2.

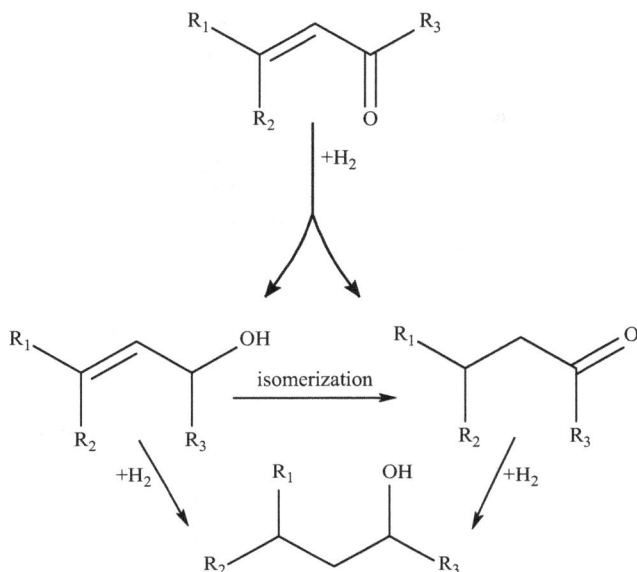

FIGURE 3.11 Hydrogenation reactions of α,β-unsaturated carbonyl group catalysed by Au/ZnO$_2$.

TABLE 3.1
Summary of Reactions Catalysed by Au NP-based Catalysts

Reaction	Nanoparticle/support	Size	Shape	Reference
Aliphatic alcohol oxidation	Au/carbon and Au/ceria	2–5 nm sized	—	[6, 32]
Oxidation of cycloalcohol	Au/CuO, Au/MnO$_2$, Au/NiO, Au/CoOx, Au/Fe$_2$O$_3$, Au/Cr$_2$O$_3$, Au/Al$_2$O$_3$, and Au/SiO$_2$	1–9 nm	—	[13]
Oxidation of aromatic alcohols	AuPd–GO/TiO$_2$ (ternary hybrid catalyst) and Au/CeO$_2$	1–7 nm	—	[20, 21]
Oxidation of amino alcohol	Au/Al$_2$O$_3$, Au/TiO$_2$, Au/MgAl$_2$O$_4$ and Au/MgO	3–4 nm sized	—	[22]
Oxidation of diols	Au/carbon	7 nm and 12 nm	—	[23]
Oxidation of polyols	Au-PGM/C and Au/charcoal, activated carbon, and Au/graphite	2–4 nm sized	—	[14, 24]

(*Continued*)

TABLE 3.1   (Continued)
Summary of Reactions Catalysed by Au NP-based Catalysts

Reaction	Nanoparticle/support	Size	Shape	Reference
Cyclohexane to adipic acid	Au on TiO_2, AlO_3, ZrO_2, MgO and CaO	2 nm (TiO_2)- 6–8 nm (CaO)	—	[18]
Epoxidation	Au/carbon	5–50 nm	—	[15]
Hydrogenation of alkenes, dienes and alkynes	Au/alumina and Au/silica	20–80 nm	—	[16, 17]
α,β-Unsaturated carbonyl groups	Au/ZnO, Au/ZrO_2	2–4 nm	Hemispherical, cuboctahedral and isohedral	[26, 27]

3.3   ORGANIC CATALYSIS BY USING AU NP

Among various noble metal NPs, Pd NPs were extensively used as catalysts because of their excellent physicochemical properties, for example, high chemical and thermal stability [8]. Along with numerous reports of nanocatalysts of metal oxide composed of diverse rare earth and transition earth metals, Pd-based nanosized catalysts have been broadly explored in academia as well as in the industry because of their applications in carbon-based cross-coupling, hydrogenation (addition reaction), esterification, oxidation and reduction reactions [33].

3.3.1   PD/RGO NP FOR REDUCTION OF NITROARENES

Pd metal NPs were synthesized on reduced graphene oxide (rGO) support in water and biological extracts. Barberry fruit extract was used to both reduce and stabilize GO and Pd^{2+} ions. These NPs were used as heterogeneous catalysts for reducing nitroarenes in presence of $NaBH_4$. The reduction of nitroarenes took place in four steps when Pd NP/rGO heterogeneous catalyst was used. The first two steps were adsorption of hydrogen on the surface of the catalyst, which was followed by adsorption of nitroarenes. The third step was an electron transfer mediated by the metal catalyst surface and the final step was the desorption of the aromatic amino compounds from the surface of the catalyst. $NaBH_4$ was used as the source of hydrogen and no reaction was observed in absence of the catalyst. Under optimized conditions, an excellent yield of 98% was obtained [34].

3.3.2   MONODISPERSE PD/GO FOR SUZUKI CROSS-COUPLING REACTION

Pd was used to catalyse many chemical reactions such as alcohol oxidation and C–C coupling reaction. Pd and homogeneous Pd complexes were important catalysts for coupling reactions because they reduced the reaction time, and resulted in high synthetic yields and selectivity. Heterogeneous Pd was used to catalyse the hydrogenation reaction. Homogeneously dispersed Pd NPs supported on graphene oxide (GO) were reported to carry out cross-coupling reactions at room temperature.

Monodisperse Pd/GO NP catalysts exhibited good activity and stability for coupling reactions. The reaction rate was faster under mild conditions and high product yields were obtained by using this catalyst. The TEM image showed that spherical Pd NPs were well-dispersed on the GO support without any signs of agglomeration [35].

3.3.3 HYDROGENATION OF UNSATURATED COMPOUNDS BY PD@GO CATALYST

Under optimized reaction conditions, *trans*-stilbene was completely reduced using Pd/GO as a catalyst in methanol (Figure 3.12).

The Pd@GO NPs exhibited better catalytic activity owing to catalysts' large surface area, ultrafine structure and the cooperative effects of the Pd with GO. It was observed that interaction between the graphene oxide and metal contributed to electron transfer. When Pd@GO catalyst is used to catalyse the Suzuki–Miyaura coupling reaction (Figure 3.13) it shows superior catalytic activity.

3.3.4 PD NP/POLYTHIOPHENE NANOSPHERES AS HETEROGENEOUS CATALYSTS

Polythiophene nanospheres were prepared by oxidative polymerization of 2-thiophenemethanol [36]. Pd NPs were thereafter decorated on these polythiophene nanospheres by a redox reaction between $PdCl_4(II)$ and 2-TPM (2-thiophene methanol) at room temperature in an aqueous solution. Under mild aerobic reaction conditions, Pd-NP/polythiophene nanospheres catalysed the Suzuki–Miyaura cross-coupling reaction with excellent activity.

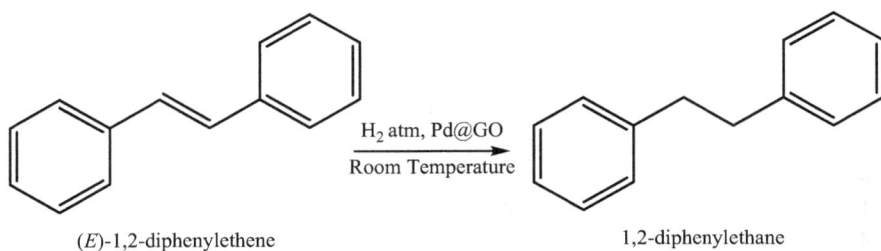

(*E*)-1,2-diphenylethene   1,2-diphenylethane

FIGURE 3.12   Hydrogenation of *trans*-stilbene on Pd/GO as catalyst.

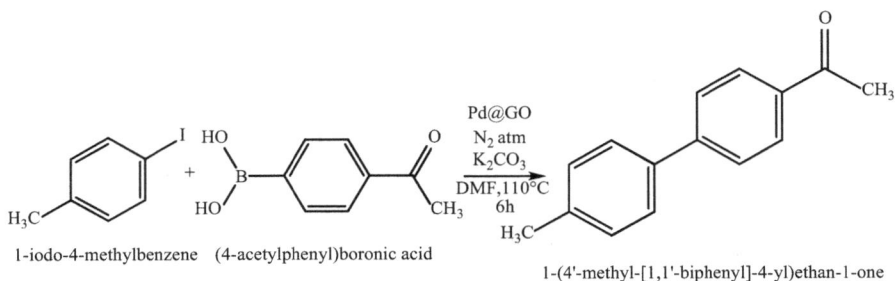

1-iodo-4-methylbenzene   (4-acetylphenyl)boronic acid

1-(4'-methyl-[1,1'-biphenyl]-4-yl)ethan-1-one

FIGURE 3.13   Suzuki–Miyaura cross-coupling using Pd/GO as catalyst.

Pd-NP/polythiophene was deposited on different substrates and evaluated as a catalyst for the Suzuki–Miyaura coupling reaction. Under mild conditions, all catalysts exhibited high conversion and selectivity. The activity of this catalyst was more than Pd/C catalysts [36].

Pd-Based catalysts were generally used in catalysing C–C bond formation reactions. The Pd(II) complexes of sterically hindered bulky and electron-rich phosphines, carbenes and palladacycles are very efficient homogeneous catalytic systems for Suzuki–Miyaura cross-coupling, which also had certain disadvantages like (1) difficulty in the isolation of product at the end of the catalytic process, (2) no reuse of catalyst in most of the cases and (3) there always remain traces of heavy metals causing contamination of the products. The heterogenization of these efficient homogeneous catalysts by immobilization or supporting on a solid surface was a logical method to overcome the difficulties faced with homogeneous catalytic systems. The use of magnetic materials as supports facilitated retrieval of the used catalyst.

The N-heterocyclic carbenes (NHCs) have been continuously dominating the world of catalysis in an unprecedented manner in the last two decades. On the other hand, Pd(II) complexes of organochalcogen ligands have emerged as efficient catalysts for C–C coupling reactions, and the air/moisture insensitive nature made them appropriate to utilize in catalytic applications. It was obvious that the NHC and the chalcogenoether (S/Se) ligands had to be mixed and the novel chalcogenoether functionalized NHCs developed in order to fine-tune and further improve the catalytic properties. The homogeneous as well as the heterogeneous version (solid-supported) of a Pd(II) complex of thioether functionalized NHC have been explored for catalytic applications. Both homogeneous and heterogeneous versions worked very efficiently in the catalysis of Suzuki–Miyaura coupling reaction under aqueous aerobic conditions. The heterogeneous version could be crafted by immobilizing the Pd(II) complex on silica-encapsulated magnetite NP (Figure 3.14).

Initially, Pd(II) complex of thioether-NHC (homogeneous version) was evaluated for catalytic activity for the Suzuki coupling reaction. During the catalytic process Pd(II) complex decomposed to afford Pd(0) species which were found to be the true catalytic species and the parent complex acted only as a pre-catalyst. The catalytic potential of these Pd(0) NPs was also investigated for Suzuki–Miyuara coupling and

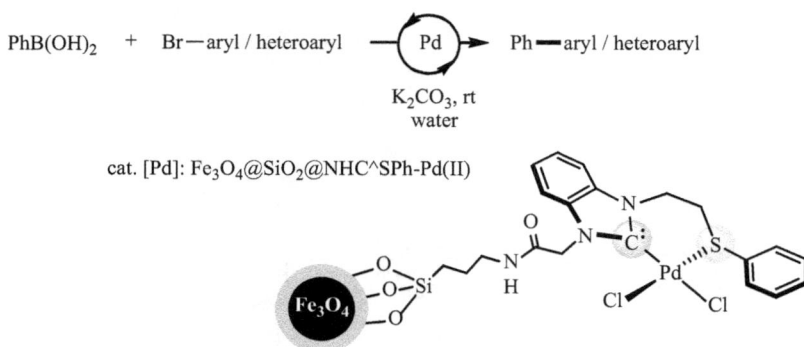

FIGURE 3.14 Suzuki–Miyaura coupling reaction catalysed with Pd(II) complex of thioether-NHC immoblized on silica-encapsulated magnetite NP.

under similar conditions, moderate yield was obtained (55%). The heterogenous version of this catalyst was obtained by immobilizing the Pd(II) complex on silica-coated Fe_3O_4 NP, which was used to overcome the problem of deactivation of the homogenous version of Pd(II) complex of thioether-NHC. The catalytic activity of composite Pd(II) catalyst was investigated for Suzuki–Miyaura coupling of aryl/heteroaryl bromides and the best results were obtained in water after 2 h at 55 °C. The magnetically active NPs were fully recovered after completion of the reaction by using an external magnet. After the isolation of the catalyst, it was reused for up to seven cycles without losing its efficiency. The unsupported homogeneous complex had a higher catalytic activity but it got deactivated after one cycle. Pd/GO were also reported as catalysts for the Suzuki–Miyaura coupling reaction [35].

The Pd(II) complex of thioether-NHC immobilized on silica-encapsulated magnetite NP (Fe_3O_4@SiO_2@NHC^SPh-Pd(II)) has also been applied for the Cu- and amine-free Sonogashira and Stille couplings of aryl bromides/chlorides under aerobic conditions. This catalyst exhibited excellent activity and was efficiently recoverable after catalysing the Sonogashira coupling under aerobic conditions and the Stille cross-coupling reactions of chloroarenes. To investigate the catalytic activity of Fe_3O_4@SiO_2@NHC^SPh-Pd(II) catalyst on the Sonogashira cross-coupling, the reaction of the acetylene and aryl bromide in DMF solvent under open-air conditions was carried out, and excellent transformation of the respective coupling products was recorded (Figure 3.15) [37].

DMF was found to be an ideal solvent for this reaction. Sonogashira reactions were usually carried out in the presence of copper as a co-catalyst and amine as a co-solvent. However, using these Pd-based catalysts, the Sonogashira coupling reaction took place in the absence of both copper and amines. It was observed that K_2CO_3 as base afforded excellent yield (96%) of the desired product when the optimum transformations produced maximum (90%) yield at 90 °C. The desired cross-coupled product was detected just after 1 h of reaction and maximum yield was obtained in 12 h. Moreover, the reaction also worked for heteroaryl bromides, as the coupling of 3-(phenylethynyl)quinoline and 3-(phenylethynyl)pyridine with phenylacetylene resulted in 69% and 75% yields of the desired products, respectively. This heterogeneous Pd(II) catalyst was an efficient recyclable catalyst that had the additional advantageous features of being filtration-free and magnetically retrievable [37].

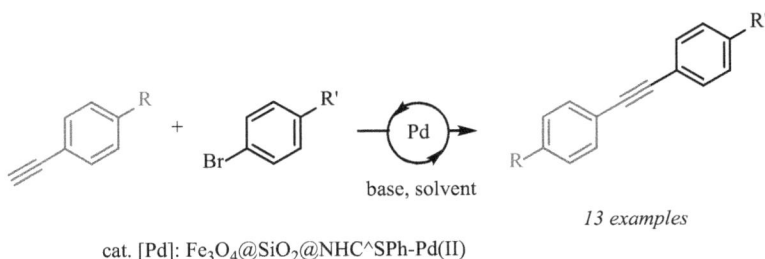

13 examples

cat. [Pd]: Fe_3O_4@SiO_2@NHC^SPh-Pd(II)

FIGURE 3.15 Sonogashira cross-coupling reaction of acetylene and aryl bromide using Fe_3O_4@SiO_2@NHC^SPh-Pd(II) as catalyst.

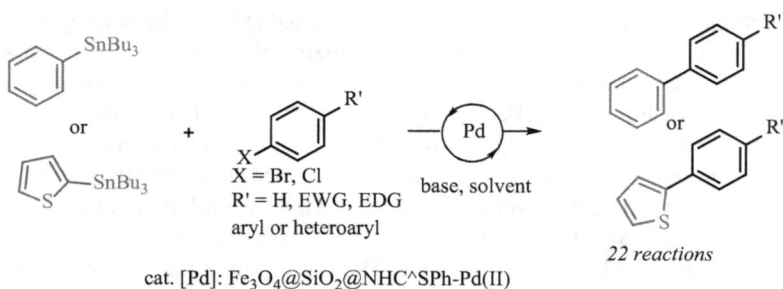

FIGURE 3.16 Stille cross-coupling reaction using Pd NP-based catalysts of tributylphenyl-stannane and aryl halides using $Fe_3O_4@SiO_2@NHC^{\wedge}SPh\text{-}Pd(II)$ as catalyst.

$Fe_3O_4@SiO_2$ supported Pd(II)-thioether-NHC catalyst showed an excellent catalytic potential on copper- /amine-free Sonogashira coupling reactions discussed above and this catalyst was also investigated for air/moisture tolerant Stille cross-coupling reaction. The Pd-catalyst worked efficiently and also activated the aryl chlorides and produced the desired coupling product under aerobic reaction conditions (Figure 3.16) [38].

To find out the optimum reaction conditions, this reaction was carried out under different reaction conditions. The satisfactory yield of the desired product was observed when water and DMF mixture was used as a solvent system and the best yield was obtained when dry DMF was used as the solvent. The base was found as an essential requirement of the reaction, where sodium acetate and potassium carbonate were found to be efficient for this coupling reaction. The working temperature of this reaction was 90 °C. The catalytic process was simple and very effective in terms of reaction time. Additionally, no fluoride salts were required, the catalyst could be recycled, and good yields of biaryls were obtained. It was the first example of the organochalcogen functionalized NHC-Pd (II) complex studied for the catalysis of Stille cross-coupling reaction. The Pd catalyst $Fe_3O_4@SiO_2@NHC^{\wedge}SPhPd(II)$ could be reused for both the Sonogashira and Stille coupling reactions [38] (Table 3.2).

TABLE 3.2
Reactions Catalysed by Pd NP-based Catalysts

Reaction	NP/Support	Size (nm)	Shape	Reference
Reduction of nitroarenes	Pd NP/reduced graphene oxide Pd@RGO	18	Spherical shaped	34
Suzuki cross-coupling reaction	Pd NP/graphene oxide Pd@GO	3.88	Spherical shaped	35
Suzuki–Miyaura cross-coupling reaction	Pd NP/polythiophene nanospheres	20		36
Suzuki–Miyaura coupling reaction	Pd-thioether NHC/Silica coated magnetite NP	1.5–2.0	Spherical shaped	38
Sonogashira coupling reaction	$Fe_3O_4@SiO_2@NHC^{\wedge}SPh\text{-}Pd(II)$			
Stille coupling reaction				

3.4 CATALYTIC REACTIONS BY IR-BASED NP

Iridium (Ir) is a rather inert element. Though the use of Ir complexes in chemical catalysis is well reported, not many applications in biological or medicinal fields exist in literature. Ultra-small Ir NPs having greater catalytic activity than their bulk counterparts were prepared by using a controlled one-step synthesis with the aid of a stabilizer that exhibited high catalytic activity under mild conditions. Ir NPs were found to be very efficient as well as selective for the catalytic reduction of nitroarenes and exhibited good potential for amino acids degradation [39]. The novel Ir catalysts demonstrated good activity for the condensation of butanol in the aqueous phase and Ir-based nanomaterials behaved as good heterogeneous catalysts for many industrial applications [10].

NPs of Ir have gained a lot of attention because of their shape- and size-dependent activity and selectivity. Additionally, tailored supports for tethering ligation sites are also of interest due to their unique catalytic properties. Among various materials, porous organic polymers have been extensively investigated due to their simple synthesis methods and ability to function as highly diverse and functional supports. Ir NPs immobilized on the surface of a porous organic polymer, triazine-phosphanimine, were used as a catalyst to reduce aldehydes to alcohols with the use of formic acid as the reducing agent. To prepare this triazine-phosphanimine porous organic polymer (TPA-POP), trichlorophosphane was reacted with melamine in the presence of dry tripropylamine which acted as a base and also as a solvent in the presence of inert gas. The reducing agent, $NaBH_4$ resulted in the formation of Ir NPs on the surface of the support. TEM analysis indicated that the size of Ir NPs was in the range of 5–10 nm [40]. The reduction of benzaldehyde with formic acid was evaluated by varying the reaction conditions and 98% yield was obtained at 80 °C with a 12 h reaction time in presence of ethyl alcohol [40].

3.4.1 IR NPS AS CATALYSTS FOR HYDROGENATION AND HYDROGEN PRODUCTION

Ir NPs exhibited excellent catalytic activity attributed to their surface, volume and quantum confinement effects. It was found that in citral, the hydrogenation of the carbonyl bond was selectively catalysed by Ir/TiO_2; but when Fe^{3+} or Ge^{4+} was added to the reaction mixture, new active sites were generated on the surface of Ir/TiO_2, which contributed towards polarizing the carbonyl bond and also enhanced the selectivity of hydrogenation of the unsaturated alcohol. The acidic or basic properties of the support governed the electronic structure of Ir. Additionally, the interaction between the Ir and the support could influence the nature of the Ir–H bond. Ir NP supported on multi-wall carbon nanotubes catalysed the p-chloronitrobenzene hydrogenation to yield p-chloroaniline at room temperature, under an H_2 atmosphere in a solvent mixture of methanol and water. Ir/Al_2O_3 was used to catalyse the decomposition of liquid hydrazine to produce N_2 and H_2. This catalytic decomposition took place in two steps. In the first step, hydrazine decomposed into NH_3 and N_2, and thereafter NH_3 decomposed into N_2 and H_2. The temperature in this exothermic reaction was found to reach approximately 1100 °C [41]. In recent years ionic liquids have been explored as electrostatic solvents which double up as capping agents. Nanoparticle synthesis in ionic liquids may be carried out from metal salts either in presence or absence of H_2 gas (reductant) and organometallic π-complexes by means of either electrodeposition or photochemical or thermal decomposition. Ir NPs were deposited on graphene to produce Ir@GO catalyst. The catalytic activity of Ir@GO composite was evaluated for the hydrogenation of benzene to cyclohexane. The Ir@TRGO nanocomposite was mixed with the substrate, benzene in an autoclave. Solvent-free reaction conditions were maintained at 100 °C under the constant pressure of 10 bar. The catalyst was recovered and could be re-used up to ten times with substantial retention of activity. The nanocomposite, under mild conditions, could catalyse hydrogenation of benzene [42].

$Ir/ZrO_2 \cdot xH_2O$ exhibited excellent activity as well as selectivity towards hydrogenation of haloaromatic nitro compounds. This catalyst displayed better activity in cyclohexane, ethanol and toluene as compared to isopropanol and THF. When a mixture of ethanol and water was used as a solvent, it substantially enhanced the activity as well as selectivity during the conversion of p-chloronitrobenzene to p-chloroaniline. The catalytic activity of Ir/ZrO_2 was much lower than that of $Ir/ZrO_2 \cdot xH_2O$. The latter worked at room temperature and atmospheric pressure and additionally there was no need for an additive or promoter. The presence of H_2O groups on $Ir/ZrO_2 \cdot xH_2O$ helped in the blocking of dehalogenation. The nitro group's electropositive N atom was easily absorbed on the surface of an electron-rich metal atom and it led to activation of N=O bond in p-chloronitrobenzene [43] (Table 3.3).

TABLE 3.3
Summary of Reactions Catalysed by Ir NP

Reaction	Nanoparticle/Support	Size (nm)	Shape	References
Reduction of aldehydes	Ir/polymeric organic support. Ir/TPA-POP	5–10	Regular in shape	40
Hydrazine decomposition	Ir/γAl₂O₃	2.1		41
Liquid phase hydrogenation of citral	Ir/TiO₂ Ir/MWCNTs			
Hydrogenation of p-chloronitrobenzene				
Hydrogenation of benzene & cyclohexene	Ir/graphene oxide composite Ir@GO	1.0–1.4		42
Hydrogenation of haloaromatic nitro compounds	Ir/hydrous zirconia Ir/ZrO₂·xH₂O	2.5–5.0		43

3.5 CATALYTIC REACTIONS BY PT-BASED NP

Nanoparticles of platinum (Pt), a rare and very expensive metal are usually used as catalysts in the form of a colloid or a suspension. Pt catalysts have applications in petroleum refining, H_2 production, automotive catalytic converters, carbon monoxide gas sensors and anticancer drugs. Their ability to catalytically oxidize CO and NO_x, electrolyse water, dehydrogenate hydrocarbons and inhibit the division of living cells has resulted in these being investigated extensively. Their size and shape were controlled by variation of the ratio of concentration of the polymer capping agent to the precursor. [44] Combination of Pt NPs with Pd resulted in a core-shell nanostructure with a dandelion-like surface that could catalyse olefin reduction reactions [45].

3.5.1 CATALYTIC APPLICATIONS OF PT NP

Supported Pt particles were used to catalyse the hydrogenation reactions of α,β-unsaturated aldehydes and alkynes at 90 °C in methanol. The catalyst was very selective for the hydrogenation of the alkynes to their corresponding alkenes. When this catalyst was used to catalyse the hydrogenation of diphenylacetylene, it was found to be highly selective to yield the *cis*-isomer of stilbene. The supported Pt NPs selectively reduced only the aldehydic group into appropriate alcohols in the case of α,β-unsaturated aldehydes, for example, only the aldehydic group of cinnamaldehyde was reduced into alcohol. For the hydrogenation of α,β-unsaturated aldehydes, the surface of Pt NPs was polarized by the magnetite nano-support imparting a partial positive charge leading to the polar functional groups of substrates being selectively adsorbed and activated. The advantage of this catalytic system was that it could be easily separated by using an external magnetic field and recycled up to four cycles with almost complete retention of activity [46].

A nanocatalyst composed of piperazine and cyanuric chloride on porous organic polymer (Pt@PC-POP) catalysed the selective hydrogenation of halogenated nitrobenzene to halogenated anilines under an atmosphere of hydrogen at room temperature. Reaction conditions were optimized using hydrogenation of *p*-chloronitrobenzene as a model reaction.

Different solvents were screened and water resulted in the conversion of *p*-chloronitrobenzene with a yield of 96.61% and a selectivity of 95.56%. When ethyl acetate was used as a solvent, the conversion was only 13.26% due to the substrate being poorly soluble in water and the catalyst having a poor dispersity in ethyl acetate. The use of ethanol or methanol as solvents resulted in *p*-chloronitrobenzene being completely hydrogenated within 1 h whereas, in ethanol, the reaction had 98% selectivity. Thus, ethanol was selected as the best solvent amongst the ones evaluated for selective catalytic hydrogenation of halogenated nitrobenzene.

Pd-based catalysts were also used to catalyse the hydrogenation of *p*-chloronitrobenzene with quantitative conversion but the selectivity was poor. In contrast, Pt-based catalysts resulted in 100% conversion and 95.48% selectivity. The general applicability of the catalyst for selective hydrogenation of halogenated nitrobenzene was also investigated. *p*-bromonitrobenzene was hydrogenated completely in *p*-bromoaniline with the selectivity of 98% in 1 h. Similarly, *m*-bromonitrobenzene was fully hydrogenated to *m*-bromoaniline with a selectivity of 97% in 1.5 h and the hydrogenation of *o*-bromonitrobenzene reached 100% conversion with 95% selectivity in 2.5 h. It was observed that for the hydrogenation of *o*-/*m*-/*p*-bromonitrobenzene the reaction time increased to 2.5 h from 1 h due to an increase in the steric hindrance of halogenated groups [47].

Honey-stabilized Pt nanostructures catalysed the formation of antipyrilquinoneimine dye by the reaction between aniline and 4-aminoantipyrine (Figure 3.17). This electron transfer reaction between aniline and 4-aminoantipyrine could be applied for the detection and removal of anilines from soil and water samples. These Pt NPs were found to accelerate the reaction between aniline and 4-aminoantipyrine. Small variations in dimensions of Pt NPs on the formation of nanowires had no detectable effect on the catalytic activity [48].

A series of supported Pt nanocatalysts over a high surface area and porous undoped and Ca-doped zinc aluminate spinels ($ZnAl_2O_4$) supports (Pt/ZCA-x, where x refers to the % of Ca) were employed for base-free glycerol oxidation. Pt NPs could be immobilized with strong affinity and were evenly distributed on the support surface. Pt/ZCA-0 catalysed the reaction with a 50.8% yield and 81% selectivity. Other oxidation products including tartonic acid, glyconic acid, etc. were also formed. In Pt/ZCA-10, Ca incorporation into spinel structure led to the increase in glycerol conversion, while a further increase of Ca in Pt/ZCA-15, resulted in a slight decrease in the catalytic activity. As compared to Pt/ZCA-0 where the Pt species were highly dispersed and had a similar density of surface basic sites, Pt/ZCA-5 showed a slight reduction in the catalytic activity. Though there was high dispersion of Pt species in the case of Pt/ZCA-15 too, its catalytic activity was found to be lower than that of Pt/ZCA-10 catalyst with the larger density of surface basic sites. Pt/ZCA-x (where x = 0, 5, 10, 15) showed very similar product selectivity, which originated from the spinel structure of ZCA-x supported resulting in similar adsorption/activation behaviour of glycerol. Pt supported by the physical mixture of ZnO, CaO and Al_2O_3, resulted in a lower activity as evidenced by a lower conversion of glycerol (16.6%) [49].

The effects of oxygen pressure and reaction temperature on glycerol oxidation over Pt/ZCA-10 were investigated. An increase in temperature to 100 °C (from 40 °C) led to a progressive increase in the rate of glycerol conversion due to the endothermic nature of the reaction. This accompanied a decrease in the selectivity towards glyceric acid at a high temperature of 80–100 °C due to further oxidation of glyceric acid and other by-products. Analogous to the reaction temperature, an increase in the oxygen pressure led to a minor decrease in the selectivity of glyceric acid.

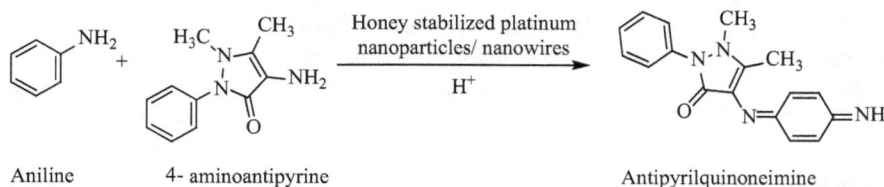

Aniline   4- aminoantipyrine   Antipyrilquinoneimine

FIGURE 3.17   The formation of antipyrilquinoneimine dye from aniline and 4-aminoantipyrine using Pt NP as catalyst.

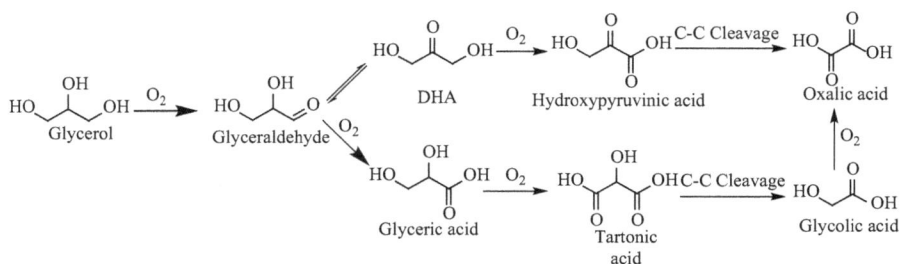

FIGURE 3.18 Oxidation process of glycerol using Pt/ZCA catalysis.

TABLE 3.4
Summary of Reactions Catalysed by Pt NP

Reaction	Nanoparticle/Support	Size (nm)	Shape	Reference
Hydrogenation reaction	Pt/magnetic NP	2–2.5	—	46
Hydrogenation of halogenated nitrobenzenes	Pt/porous organic polymer Pt@PC-POP	2.96	—	47
Oxidation reaction	Pt/Ca-doped ZnAl$_2$O$_4$ spinels	3.13	—	49

Glycerol could be easily oxidized by the Pt/ZCA to glyceric acid, which was accompanied by the conversion of glycerol to DHA. DHA and glyceric acid were formed through different pathways (Figure 3.18). Glyceraldehyde, a reaction intermediate could not be detected because of its fast oxidation to glyceric acid [49] (Table 3.4).

3.6 CATALYTIC REACTIONS BY AG-BASED NP

Among various metal NPs, silver nanoparticles (Ag NPs) are of great interest in organic catalysis because nano-Ag catalysts have distinctive chemical reactivity, stability and excellent selectivity along with high recyclability [12]. Ag NPs also have an interesting optical property, the localized surface plasmon resonance (LSPR), which is key to the recent developments in catalysis for both oxidation and reduction [50]. Nanocatalysis processes could be considered as a bridging methodology between heterogeneous and homogeneous catalysis. Most of the NPs behave as the heterogeneous catalyst during catalysis and it has several advantages like being ligand-free, and recyclable, facilitating low temperature and faster reactions that are efficient and selective [51].

3.6.1 APPLICATION OF AG NP

Ag NPs played a significant role in the catalysis taking part in many organic transformations in the synthesis of fine chemicals. Ag NPs were vastly used for reducing molecules with different functionality, such as nitroaromatics and carbonyl compounds [12].

3.6.1.1 Reduction of Nitroaromatics

Ag NPs were reported as catalysts for the reduction of nitroaromatics. 4-Nitrophenol was reduced in water using a simple process that helped in evaluating the catalytic activity of Ag NP. The Ag NP on different kinds of supports such as Ag nano shell-coated cationic polystyrene beads, Ag nanotubes, nano-Ag/silica nanotubes, etc. form efficient catalysts for water purification. Some catalysts like Ag hydrogel-based nanocatalyst, bimetallic Au, Ag core-shell nanoparticle, etc. were designed to reduce 4-nitrophenol [12].

Selective reduction of nitroaromatics was carried out by support-dependent Ag cluster, which afforded selective reduction. The Ag NP on Al_2O_3 was found to be highly efficient catalysts for the hydrogenation of 4-nitrostyrene. Reduction of a series of nitroaromatic compounds was carried out selectively leading to the hydrogenation of the nitro group to amine with quantitative conversions [12].

It was observed that the Ag NPs on different supports, like nanoporous alumina membranes and polystyrene nanotubes, exhibited higher catalytic activity than Ag NPs without support. Cross-linked lysozyme crystals (CLLC) proved to be a novel bio-nano composite support material for Ag NPs, with the resulting nano-Ag-based catalyst showing high catalytic activity for the reduction of nitroaromatic compounds, including 4-nitrophenol. The nano-Ag catalyst in the presence of intermolecular or intramolecular olefinic functionalities showed complete chemoselectivity for the nitro groups. Also, the nano-Ag catalyst was found to be reusable without any significant loss of catalytic activity or selectivity. Ag@Ni magnetic nanocatalysts were used for the hydrogenation of nitroaromatic compounds. This catalyst could be reused up to eight times without any significant loss of yield of the desired product [12]. Sometimes, the reduction of nitroaromatic compounds produced intermediate compounds due to partial hydrogenation [12].

3.6.1.2 Reduction of Carbonyl Compounds

The catalytic activity of Ag NPs towards the alkene or carbonyl functional groups was quite low. However, the core-shell Ag@Ni nanocatalysts were very active for the reduction of carbonyl compounds (Figure 3.19) [52]. On reduction reaction of carbonyls, corresponding alcohols were produced with excellent yields under the hydrogenation conditions [12].

Use of core-shell AgNP@CeO_2 dispersed on CeO_2 support was used for reduction of substrates like unsaturated aldehydes and alkyl aldehydes with olefin functionality (Figure 3.20) [53]. This catalyst was easily recovered by simple filtration and no leaching of Ag was observed. Also, this catalyst was found to be suitable for reuse without any loss of its high efficiency [12].

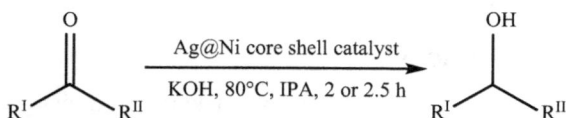

FIGURE 3.19   Reduction of ketone to secondary alcohol using Ag@Ni nanocatalyst.

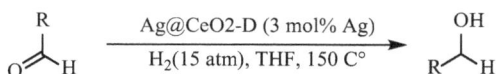

FIGURE 3.20 Reduction of aldehyde to primary alcohol using AgNP@CeO$_2$ catalyst.

3.6.1.3 Other Reductive Transformations

Ag NP attached on hydrotalcite behaved as a catalyst using alcohol as a reductant for deoxygenation of different epoxides into alkenes. This catalyst was found to be highly effective for the deoxygenation process of epoxides having phenyl groups like styrene oxide but not effective for the deoxygenation of aliphatic epoxides (Figure 3.21) [12].

Ag-ceria core-shell nanocomposite catalyst was generally used to catalyse the reduction of nitroaromatic compounds. AgNP@CeO$_2$ catalyst had an exclusive reduction ability for deoxygenation of epoxides and converted them into their corresponding alkenes. Epoxides of aromatic, alicyclic and aliphatic compounds were converted into the corresponding alkenes with high yields and selectivity. It was also found that this catalyst could be recovered by using a facile filtration method and could be reused with substantial retention of activity and selectivity [12].

3.6.1.4 Oxidation of Alcohols

Oxidation reaction is a fundamental process in nature and a very important transformation in organic chemistry. Various nano-Ag catalyst systems were determined as effective heterogeneous catalysts for the oxidation of alcohols. Selective oxidation reaction of alcohols to carbonyl compounds or their derivatives has been one of the most significant and fundamental reactions of organic chemistry [12]. It was found that the monodispersed Au-Ag alloy clusters with varying Ag content were used for catalytic oxidation of p-hydroxybenzyl alcohol [54]. It was also observed that the presence of less than 10% Ag on Au–Ag NP enhanced the catalytic activity of the catalyst. The gas-phase oxidation of high–molecular-weight alcohol to corresponding aliphatic aldehyde was carried out by a novel catalyst of Ag NP fixed in a silicon nanowire (Ag@SiNW) [55]. The Ag@SiNW catalyst showed higher selectivity than other Ag catalysts towards the desired product for conversion of octan-1-ol. Ag NP-supported hydrotalcite catalyst showed very high catalytic activity for dehydrogenated oxidation of a variety of molecules containing the alcohol functional group.

3.6.1.5 Oxidation of Silanes

Silanes and silanols were useful in the synthesis of organosilicon-based polymeric materials. It was observed that the Ag NP with hydroxyapatite base showed high

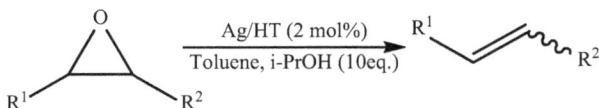

FIGURE 3.21 Catalytic deoxygenation of epoxide to alkene using Ag/HT catalyst.

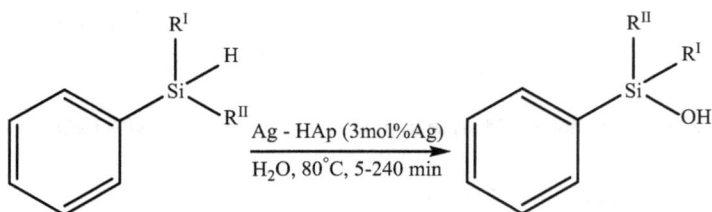

FIGURE 3.22 Oxidation of silanes to silanols using Ag-HAp as catalyst.

catalytic activity for selective oxidation of hydrosilanes, where the aqueous medium was used as the green oxidant (Figure 3.22). Moreover, Ag-hydroxyapatite (Ag-HAp) catalyst could be recovered from the reaction mixture and could be easily reused up to four times without losing catalytic efficiency and selectivity. The catalytic activity of Ag-based NP was investigated and it was found that both aromatic and aliphatic silanes could be effectively transformed into the corresponding silanols and silyl ethers, respectively by using alcohols and water as oxidants (Figure 3.23) [12].

Recently, halloysite nanotube-supported Ag NPs were found to behave as an active catalyst for carrying out the polymerization of an alkylsilane ($C_{18}H_{37}SiH_3$) in an aqueous solution to form silanol- /siloxane-based composite microspheres [12].

3.6.1.6 Oxidation of Olefins

Olefinic substrates are one important class of organic intermediates, due to their easy availability and possibility to get transformed into various functional compounds. Ag NPs were generally used to catalyse the oxidation of olefins and aromatics [55]. In 2006, Shape-tailored Ag NP catalysts helped to catalyse the oxidation of styrene but the results of this reaction were not good. Later, Purcar and co-workers prepared a novel nano-Ag catalyst supported on the hybrid films, which contained methyltrime-thoxysilane (Ag-MeTMS) to catalyse the oxidation of styrene, and excellent catalytic

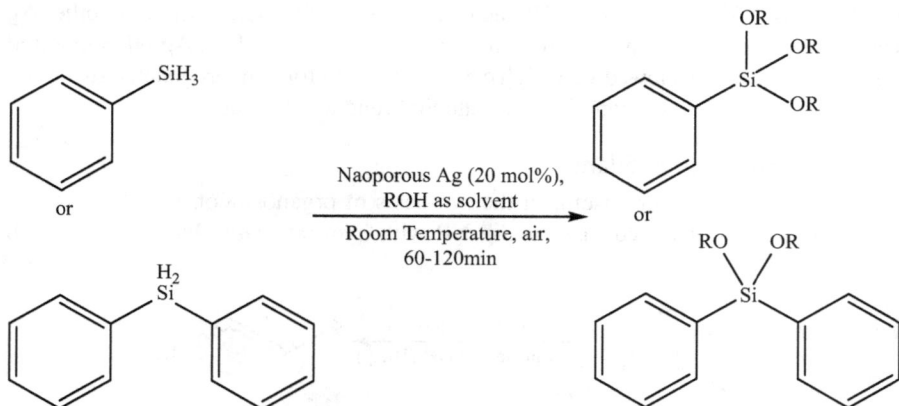

FIGURE 3.23 Oxidation of silanes with alcohol to silyl ether using Ag-based NP.

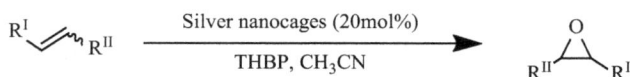

FIGURE 3.24   Nano-Ag catalysed oxidation of alkene.

activity was observed with ~99% yield. It was also reported that the Ag nanocages helped in the oxidation of variety of olefins (Figure 3.24). Acyclic and cyclic olefins formed corresponding epoxides with high selectivity and high conversion rate [12].

3.6.1.7   Alkylation of Amines and Arenes

Alkylation is of great interest to organic chemists for synthesizing a large number of complex chemicals and biologically active molecules. Silica-supported Ag NPs helped to catalyse the Friedel–Crafts alkylation of arenes (Figure 3.25). The catalyst (Ag_5SiO_2) was active towards the Friedel–Crafts alkylation of arenes, whereas it showed lower reactivity for aliphatic alcohols, styrene and indene. The Ag NP was easily recovered from the reaction mixture by a simple filtration and could be reused. But the chemoselectivity of the product was poor due to the formation of the mixture *o-/p*-substituted aromatics [12].

3.6.1.8   Miscellaneous Applications

3.6.1.8.1   Ag NP-catalysed Green Synthesis and Reduction of Methylene Blue Dye

Synthetic organic dyes used in numerous industries are harmful to health and environment. Ag nano-catalysts synthesized by green methodology were used for reduction of methylene blue dye in the presence of sodium borohydride ($NaBH_4$) [56].

3.6.2   Magnetic NP

The separation of solid heterogeneous catalysts from a reaction mixture could be carried out by the filtration or centrifugation process. Catalysts with nanometer-sized

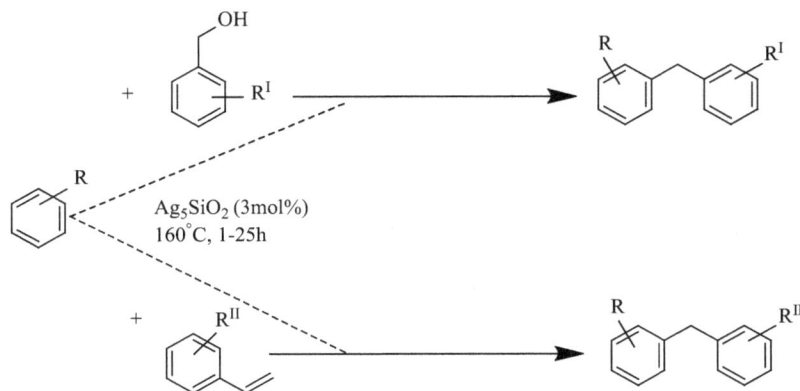

FIGURE 3.25   Catalytic Friedel–Crafts alkylation of arenes.

magnetic core and catalytic shell were used as magnetically recyclable catalysts and could be recycled conveniently by the application of an external magnetic field. For example, Ag and maghemite (γ-Fe_2O_3) NPs were used to construct magnetically recyclable nanometer-sized catalysts that were used to catalyse the reduction of 4-nitrophenol in presence of $NaBH_4$. Spongy type Au and Ag nanostructures were reported for the reduction of 4-nitrophenol. These nanostructures possessed high catalytic activity, but during the use of these nanostructures, it was difficult to avoid their contamination caused by the formation of undesired by-products in the reaction medium. The Fe_2O_3 NPs deposited on Ag were found to be efficient nanocatalysts for carrying out the reduction reaction, which converted 4-nitrophenol to 4-aminophenol by using $NaBH_4$ in the aqueous media. The magnetic particles were harnessed easily from the solution by using a neodymium magnet. The catalysts could be recycled several times [33] (Table 3.5).

3.6.2.1   Future Perspectives

In recent times, metal NPs have emerged as a prominent class of catalysts that are capable of converting the organic compounds into desired products with high yield and high stereoselectivity. With the rapid development of noble metal nanocatalysts and their demonstrated application in synthetic organic chemistry, researchers working in this field are exploring this vigorously. This book chapter has compiled and illustrated all major strategies utilized for the synthesis of simple and biologically significant organic molecules catalysed by Au, Pd, Ir, Pt and Ag NPs. These noble

TABLE 3.5
Summary of Reactions Catalysed by Ag NP

Reaction	Nanoparticle/Support	Size (nm)	Reference
Reduction reaction	Ag NP/polystyrene beads	—	[12, 37, 52–55]
Hydrogenation of nitroaromatics	Ag Nanocluster/ Θ-Al_2O_3	0.8–4.2	
Reduction of carbonyl compounds	Ag/Ni magnetic Ag@Ni magnetic nanocatalyst	—	
Reduction of unsaturated aldehydes	Ag/CeO_2 Ag NP@CeO_2	—	
Deoxygenation of epoxides	Ag NP/hydrotalcite	—	
Oxidation of alcohols	Ag/silicon nanowire	—	
Oxidation of hydrosilanes	Ag/hydroxyapatite Ag-Hap	—	
Oxidation of organosilanes	Ag nanoporous (No support)	25	
Epoxidation reaction	Ag NP cages	5 ± 2	
Alkylation of arenes	Ag NP/silica		
Reduction of methylene	Ag NP (no support)	8–32	[56]
Reduction of 4-nitrophenol	Ag/magnetic NP	5–10	[57]

metal NPs are seen to have great potential to be developed as high-performance catalysts which will become an advantageous tool for synthetic organic chemists. This in turn would positively influence a plethora of fields like pharmaceuticals, petrochemicals, energy conversion to name a few. We wish that this book chapter will provide useful insight on the chemical transformations catalysed by noble metal NPs and encourage students to further study this as well as researchers to take up new challenges for further development of this field.

REFERENCES

1. Navalón, S. & García, H., (2016). Nanoparticles for catalysis. *Nanomaterials*, 6 (7), 123.
2. Jeevanandam, J., Barhoum, A., Chan, Y. S., Dufresne, A. & Danquah, M. K., (2018). Review on nanoparticles and nanostructured materials: history, sources, toxicity and regulations. *Beilstein Journal of Nanotechnology*, 9 (1), 1050–1074.
3. Rodrigues, T., Marques da Silva, A. & Camargo, P. (2019). Nanocatalysis by Noble metal nanoparticles: controlled synthesis for the optimization and understanding of activities. *Journal of Materials Chemistry A*, 7(11), 5857–5874.
4. Gawande, M. B., Goswami, A., Felpin, F.-X., Asefa, T., Huang, X., Silva, R., Zou, X., Zboril, R. & Varma, R. S., (2016). Cu and Cu-based nanoparticles: synthesis and applications in catalysis. *Chemical Reviews*, 116 (6), 3722–3811. doi:10.1021/acs.chemrev.5b00482
5. Kharissova, O. V., Dias, H. V. R., Kharisov, B. I., Pérez, B. O. & Pérez, V. M. J., (2013). The greener synthesis of nanoparticles. *Trends in Biotechnology*, 31 (4), 240–248. https://www.sciencedirect.com/science/article/pii/S0167779913000152
6. Pflästerer, D. & Hashmi, A. S. K., (2016). Gold catalysis in total synthesis – recent achievements. *Chemical Society Reviews*, 45 (5), 1331–1367. doi:10.1039/C5CS00721F
7. Hashmi, A. S. K., (2007). Gold-catalyzed organic reactions. *Chemical Reviews*, 107 (7), 3180–3211. doi:10.1021/cr000436x
8. Chen, A. & Ostrom, C., (2015). Palladium-based nanomaterials: synthesis and electrochemical applications. *Chemical Reviews*, 115 (21), 11999–12044. doi:10.1021/acs.chemrev.5b00324
9. Cui, M., Huang, X., Zhang, X., Xie, Q. & Yang, D., (2020). Ultra-small iridium nanoparticles as active catalysts for the selective and efficient reduction of nitroarenes. *New Journal of Chemistry*, 44 (42), 18274–18280. doi:10.1039/D0NJ03621H
10. Liu, D., Chen, X., Xu, G., Guan, J., Cao, Q., Dong, B., Qi, Y., Li, C. & Mu, X., (2016). Iridium nanoparticles supported on hierarchical porous N-doped carbon: an efficient water-tolerant catalyst for bio-alcohol condensation in water. *Scientific Reports*, 6 (1), 1–13.
11. Saternus, M., Fornalczyk, A., Gąsior, W., Dębski, A. & Terlicka, S., (2020). Modifications and improvements to the collector metal method using an mhd pump for recovering platinum from used car catalysts. *Catalysts*, 10 (8), 880.
12. Dong, X.-Y., Gao, Z.-W., Yang, K.-F., Zhang, W.-Q. & Xu, L.-W., (2015). Nanosilver as a new generation of silver catalysts in organic transformations for efficient synthesis of fine chemicals. *Catalysis Science & Technology*, 5 (5), 2554–2574. doi:10.1039/C5CY00285K
13. Hutchings, G., Carrettin, S., Landon, P., Edwards, J., Enache, D., Knight, D., Xu, Y.-J. & Carley, A., (2006). New approaches to designing selective oxidation catalysts: Au/C a versatile catalyst. *ChemInform*, 38, 223–230.
14. Carabineiro, S. A. C., (2019). Supported gold nanoparticles as catalysts for the oxidation of alcohols and alkanes. *Frontiers in Chemistry*, 7, 702–702. https://www.frontiersin.org/article/10.3389/fchem.2019.00702

15. Dimitratos, N., Porta, F., Prati, L. & Villa, A., (2005). Synergetic effect of platinum or palladium on gold catalyst in the selective oxidation of D-sorbitol. *Catalysis Letters*, 99, 181–185.

16. Dimitratos, N. & Prati, L., (2005). Gold based bimetallic catalysts for liquid phase applications. *Gold Bulletin*, 38 (2), 73–77. doi:10.1007/BF03215236

17. Carrettin, S., McMorn, P., Johnston, P., Griffin, K. & Hutchings, G. J., (2002). Selective oxidation of glycerol to glyceric acid using a gold catalyst in aqueous sodium hydroxide. *Chemical Communications*, (7), 696–697. doi:10.1039/B201112N

18. Biella, S., Castiglioni, G. L., Fumagalli, C., Prati, L. & Rossi, M., (2002). Application of gold catalysts to selective liquid phase oxidation. *Catalysis Today*, 72 (1), 43–49. https://www.sciencedirect.com/science/article/pii/S092058610100476X

19. Villa, A., Campisi, S., Schiavoni, M. & Prati, L., (2013). Amino alcohol oxidation with gold catalysts: the effect of amino groups. *Materials*, 6 (7), 2777–2788.

20. Wang, H., Fan, W., He, Y., Wang, J., Kondo, J. & Tatsumi, T., (2013). Selective oxidation of alcohols to aldehydes/ketones over copper oxide-supported gold catalysts. *Journal of Catalysis*, 299, 10–19.

21. Abad, A., Almela, C., Corma, A. & García, H., (2006). Efficient chemoselective alcohol oxidation using oxygen as oxidant. Superior performance of gold over palladium catalysts. *Tetrahedron*, 62 (28), 6666–6672. https://www.sciencedirect.com/science/article/pii/S0040402006007058

22. Abad, A., Almela, C., Corma, A. & García, H., (2006). Unique gold chemoselectivity for the aerobic oxidation of allylic alcohols. *Chemical Communications*, (30), 3178–3180. doi:10.1039/B606257A

23. Abad, A., Corma, A. & García, H., (2008). Catalyst parameters determining activity and selectivity of supported gold nanoparticles for the aerobic oxidation of alcohols: the molecular reaction mechanism. *Chemistry – A European Journal*, 14 (1), 212–222. doi:10.1002/chem.200701263

24. Dias Ribeiro de Sousa Martins, L. M., Carabineiro, S. A. C., Wang, J., Rocha, B. G. M., Maldonado-Hódar, F. J. & Latourrette de Oliveira Pombeiro, A. J., (2017). Supported gold nanoparticles as reusable catalysts for oxidation reactions of industrial significance. *ChemCatChem*, 9 (7), 1211–1221. doi: 10.1002/cctc.201601442

25. Wang, J., Kondrat, S. A., Wang, Y., Brett, G. L., Giles, C., Bartley, J. K., Lu, L., Liu, Q., Kiely, C. J. & Hutchings, G. J., (2015). Au–Pd nanoparticles dispersed on composite titania/graphene oxide-supports as a highly active oxidation catalyst. *ACS Catalysis*, 5 (6), 3575–3587. doi:10.1021/acscatal.5b00480

26. Alshammari, A. & Kalevaru, V. N., (2016). Supported gold nanoparticles as promising catalysts. *Catalytic Application of Nano-Gold Catalysts*. IntechOpen, London, United Kingdom. 57–81.

27. Hughes, M. D., Xu, Y.-J., Jenkins, P., McMorn, P., Landon, P., Enache, D. I., Carley, A. F., Attard, G. A., Hutchings, G. J., King, F., Stitt, E. H., Johnston, P., Griffin, K. & Kiely, C. J., (2005). Tunable gold catalysts for selective hydrocarbon oxidation under mild conditions. *Nature*, 437 (7062), 1132–1135. http://europepmc.org/abstract/MED/16237439. doi:10.1038/nature04190

28. Erkelens, J., Kemball, C. & Galwey, A. K., (1963). Some reactions of cyclohexene with hydrogen and deuterium on evaporated gold films. *Transactions of the Faraday Society*, 59 (0), 1181–1191. doi:10.1039/TF9635901181

29. Bond, G. C., Sermon, P. A., Webb, G., Buchanan, D. A. & Wells, P. B., (1973). Hydrogenation over supported gold catalysts. *Journal of the Chemical Society, Chemical Communications*, (13), 444b–445. doi:10.1039/C3973000444B

30. Bailie, J. E. & Hutchings, G. J., (1999). Promotion by sulfur of gold catalysts for crotyl alcohol formation from crotonaldehyde hydrogenation. *Chemical Communications*, (21), 2151–2152. doi:10.1039/A906538E

31. Mohr, C., Hofmeister, H., Radnik, J. & Claus, P., (2003). Identification of active sites in gold-catalyzed hydrogenation of acrolein. *Journal of the American Chemical Society*, 125 (7), 1905–1911. doi:10.1021/ja027321q

32. Debecker, D. P. & Mutin, P. H., (2012). Non-hydrolytic sol–gel routes to heterogeneous catalysts. *Chemical Society Reviews*, 41 (9), 3624–3650. doi:10.1039/C2CS15330K

33. Shakil Hussain, S. M., Kamal, M. S. & Hossain, M. K., (2019). Recent developments in nanostructured palladium and other metal catalysts for organic transformation. *Journal of Nanomaterials*, 2019, 1562130–1562130. doi:10.1155/2019/1562130

34. Nasrollahzadeh, M., Sajadi, S. M., Rostami-Vartooni, A., Alizadeh, M. & Bagherzadeh, M., (2016). Green synthesis of the Pd nanoparticles supported on reduced graphene oxide using barberry fruit extract and its application as a recyclable and heterogeneous catalyst for the reduction of nitroarenes. *Journal of Colloid and Interface Science*, 466, 360–368. http://europepmc.org/abstract/MED/26752431. doi:10.1016/j.jcis.2015.12.036

35. Diler, F., Burhan, H., Genc, H., Kuyuldar, E., Zengin, M., Cellat, K. & Sen, F., (2020). Efficient preparation and application of monodisperse palladium loaded graphene oxide as a reusable and effective heterogeneous catalyst for Suzuki cross-coupling reaction. *Journal of Molecular Liquids*, 298, 111967–111967. https://www.sciencedirect.com/science/article/pii/S0167732219353887

36. Bae, S.-E., Kim, K.-J., Hwang, Y.-K. & Huh, S., (2015). Simple preparation of Pd-NP/polythiophene nanospheres for heterogeneous catalysis. *Journal of Colloid and Interface Science*, 456, 93–99.

37. Nayan Sharma, K., Satrawala, N. & Kumar Joshi, R., (2018). Thioether–NHC-ligated PdII complex for crafting a filtration-free magnetically retrievable catalyst for Suzuki–Miyaura coupling in water. *European Journal of Inorganic Chemistry*, 2018 (16), 1743–1751. doi:10.1002/ejic.201800209

38. Khandaka, H., Sharma, K. N. & Joshi, R. K., (2021). Aerobic Cu and amine free Sonogashira and Stille couplings of aryl bromides/chlorides with a magnetically recoverable $Fe_3O_4@SiO_2$ immobilized Pd(II)-thioether containing NHC. *Tetrahedron Letters*, 67, 152844–152844. https://www.sciencedirect.com/science/article/pii/S0040403921000241

39. Goel, A. & Lasyal, R., (2016). Iridium nanoparticles with high catalytic activity in degradation of acid red-26: an oxidative approach. *Water Science and Technology*, 74 (11), 2551–2559. doi:10.2166/wst.2016.330

40. Panahi, F., Haghighi, F. & Khalafi-Nezhad, A., (2020). Reduction of aldehydes with formic acid in ethanol using immobilized iridium nanoparticles on a triazine-phosphanimine polymeric organic support. *Applied Organometallic Chemistry*, 34 (10), e5880–e5880. doi:10.1002/aoc.5880

41. Cui, M.-L., Chen, Y.-S., Xie, Q.-F., Yang, D.-P. & Han, M.-Y., (2019). Synthesis, properties and applications of noble metal iridium nanomaterials. *Coordination Chemistry Reviews*, 387, 450–462. https://www.sciencedirect.com/science/article/pii/S0010854517305696

42. Marcos Esteban, R., Schütte, K., Brandt, P., Marquardt, D., Meyer, H., Beckert, F., Mülhaupt, R. & Kölling, H., (2015). Iridium@graphene composite nanomaterials synthesized in ionic liquid as re-usable catalysts for solvent-free hydrogenation of benzene and cyclohexene. *Nano-Structures & Nano-Objects*, 2, 11–18.

43. Fan, G.-Y., Zhang, L., Fu, H.-Y., Yuan, M.-L., Li, R.-X., Chen, H. & Li, X.-J., (2010). Hydrous zirconia supported iridium nanoparticles: An excellent catalyst for the hydrogenation of haloaromatic nitro compounds. *Catalysis Communications*, 11 (5), 451–455. https://www.sciencedirect.com/science/article/pii/S1566736709004439

44. Forbes, L. M., (2013). *Controlling the Growth and Catalytic Activity of Platinum Nanoparticles Using Peptide and Polymer Ligands.* University of California, San Diego.
45. Datta, K. J., Datta, K. K. R., Gawande, M. B., Ranc, V., Čépe, K., Malgras, V., Yamauchi, Y., Varma, R. S. & Zboril, R., (2016). Pd@Pt core–shell nanoparticles with branched dandelion-like morphology as highly efficient catalysts for olefin reduction. *Chemistry – A European Journal*, 22 (5), 1577–1581. doi:10.1002/chem.201503441
46. Abu-Reziq, R., Wang, D., Post, M. & Alper, H., (2007). Platinum Nanoparticles Supported on Ionic Liquid-Modified Magnetic Nanoparticles: Selective Hydrogenation Catalysts. *Advanced Synthesis & Catalysis*, 349 (13), 2145–2150. doi:10.1002/adsc.200700129
47. Zhao, H., Yu, G., Yuan, M., Yang, J., Xu, D. & Dong, Z., (2018). Ultrafine and highly dispersed platinum nanoparticles confined in a triazinyl-containing porous organic polymer for catalytic applications. *Nanoscale*, 10 (45), 21466–21474. doi:10.1039/C8NR05756G
48. Venu, R., Ramulu, T. S., Anandakumar, S., Rani, V. S. & Kim, C. G., (2011). Bio-directed synthesis of platinum nanoparticles using aqueous honey solutions and their catalytic applications. *Colloids and Surfaces A: Physicochemical and Engineering Aspects*, 384 (1), 733–738. https://www.sciencedirect.com/science/article/pii/S0927775711003724
49. Han, Z., Xie, R., Song, Y., Fan, G., Yang, L. & Li, F., (2019). Efficient and stable platinum nanocatalysts supported over Ca-doped ZnAl$_2$O$_4$ spinels for base-free selective oxidation of glycerol to glyceric acid. *Molecular Catalysis*, 477, 110559–110559. https://www.sciencedirect.com/science/article/pii/S2468823119304043
50. Li, A. Y., Gellé, A., Segalla, A. & Moores, A. (2019). *Silver Nanoparticles in Organic Transformations*. Wiley-VCH, Germany 723–793.
51. Bhanage, M. A. B. & Bhalchandra, M., (2015). Silver nanoparticles: synthesis, characterization and their application as a sustainable catalyst for organic transformations. *Current Organic Chemistry*, 19 (8), 708–727. http://www.eurekaselect.com/node/128283/article
52. Gawande, M. B., Guo, H., Rathi, A. K., Branco, P. S., Chen, Y., Varma, R. S. & Peng, D.-L., (2013). First application of core-shell Ag@Ni magnetic nanocatalyst for transfer hydrogenation reactions of aromatic nitro and carbonyl compounds. *RSC Advances*, 3 (4), 1050–1054. doi:10.1039/C2RA22143H
53. Mitsudome, T., Mikami, Y., Matoba, M., Mizugaki, T., Jitsukawa, K. & Kaneda, K., (2012). Design of a silver–cerium dioxide core–shell nanocomposite catalyst for chemoselective reduction reactions. *Angewandte Chemie International Edition*, 51 (1), 136–139. doi:10.1002/anie.201106244
54. Chaki, N. K., Tsunoyama, H., Negishi, Y., Sakurai, H. & Tsukuda, T., (2007). Effect of Ag-doping on the catalytic activity of polymer-stabilized Au clusters in aerobic oxidation of alcohol. *The Journal of Physical Chemistry C*, 111 (13), 4885–4888. doi:10.1021/jp070791s
55. Crites, C.-O. L., Hallett-Tapley, G. L., Frenette, M., González-Béjar, M., Netto-Ferreira, J. C. & Scaiano, J. C., (2013). Insights into the mechanism of Cumene peroxidation using supported gold and silver nanoparticles. *ACS Catalysis*, 3 (9), 2062–2071. doi:10.1021/cs400337w
56. Saha, J., Begum, A., Mukherjee, A. & Kumar, S., (2017). A novel green synthesis of silver nanoparticles and their catalytic action in reduction of methylene blue dye. *Sustainable Environment Research*, 27 (5), 245–250. https://www.sciencedirect.com/science/article/pii/S2468203916302801
57. Shin, K., Choi, J.-Y., Park, C. S., Jang, H. & Kim, K., (2009). Facile synthesis and catalytic application of silver-deposited magnetic nanoparticles. *Catalysis Letters*, 133, 1–7.

4 Organometallic Compounds as Heterogeneous Catalysts

Garima Sachdeva
Amity University Haryana, Gurugram (Haryana), India

Monu Verma
The University of Seoul, Seoul, Republic of Korea

Varun Rawat
Amity University Haryana, Gurugram (Haryana), India

Ved Prakash Verma, Manish Srivastava, Sudesh Kumar, and Vanshika Singh
Banasthali Vidyapith, Jaipur, India

CONTENTS

DOI: 10.1201/9781003126270-4

4.1   INTRODUCTION

Heterogeneous catalysis plays a significant role in industrial processes these days as it helps to obtain selective and reusable catalysts. At first glance, heterogeneous catalysis looks very different from organometallic chemistry and homogeneous catalysis. The molecular approach is now widespread in heterogeneous catalysis, and continuity of disciplines is being accepted that runs from monometallic activation to solid-state activation through organometallic clusters. Organometallic chemistry is an exciting and active field of research with several practical applications which involve compounds with at least one metal–carbon bond [1]. It links the aspects of both organic as well as inorganic chemistry, hence is of paramount importance. Studies of organometallic compounds have improved the understanding of chemical bonding because these complexes have distinct structures and bonds. The field of organometallic chemistry originated in the mid-1800s when Frankland found the ethyl and methyl derivatives of Zinc, Tin, and Mercury, and is still a growing field of interest. The reason behind the rapid growth of organometallic chemistry is due to the wide-ranging applications of organometallic complexes in organic synthesis (which was noticed with the discovery of the Grignard reagent at the end of the 19th century) and the function of metals in biological systems [2–4]. Organometallic catalysts can catalyze many organic reactions due to the ease of modification in their environment by changing the surrounding ligands. Many ligands can coordinate with the transition metals, which are responsible for changing the reactivity and selectivity completely. Few organometallic reagents are extremely specific, which allows the preparation of the complex target without involving the protecting groups. Complex molecules can be combined with the help of organometallic reagents, which provides flexibility and is effective in designing functional materials and biologically important molecules [5]. Organometallic compounds are also utilized as precursors in the preparation of nanomaterials and microelectronic materials. As organometallic compounds consist of ligands and metals, the synthetic methods can be divided into two types: the reaction between metal species and ligand precursor, and the reaction of the organometallic compound to yield a new ligand.

The organometallic chemistry field is further divided into two domains based on the metal bonded with carbon. One includes the compounds having alkali, alkaline earth metals, and more metallic elements in the Zn, B, C, O, and N groups of the periodic table. Carbon is bonded by a simple σ-bond or by the ionic bond in these compounds. Examples of such compounds include Grignard reagents, organolithium reagents, etc. While the other involves the compounds consisting of transition metals like ferrocene or Ziese's salt which have a metal–carbon π-bond [6]. Vitamin B_{12} co-enzyme, which includes a Co–C linkage, is a prominent example of an organometallic complex.

The properties of organometallic compounds differ on the basis of bonds involved between metal and carbon. This distinction can be used for different applications like the production of commodity chemicals (such as polymers), medicinal chemistry, fine chemical synthesis, and catalysis [1, 7]. Organometallic compounds which possess the properties of both ionic and covalent bonds play a crucial role in industrial chemical reactions as they are stable in solutions and sufficiently ionic to undergo the reactions [8]. Organometallic compounds can work in homogeneous as well as heterogeneous catalysis, but we will be discussing the organometallic compounds in heterogeneous catalysis.

4.2   ORGANOCOPPER COMPOUNDS

Copper is the cheapest and most abundant of the coinage metals, with oxidation states ranging from 0 to +4 [9]. Tremendous advancement has been seen in organic synthesis with copper as a reagent and catalyst in the last few years. Organocopper compounds consist of a carbon–copper chemical bond and provide effective coupling of two different carbon groups. Organocopper reagents are easy to handle, highly reactive, and provide regioselectivity, chemoselectivity, and stereoselectivity; hence, they are utilized in the synthesis of natural products and as intermediates for many reactions.

One of the most commonly used methods for the preparation of organocopper compounds is the transmetalation method. The transmetalation reaction is one of the earliest and simplest reactions that make up the core of organometallic chemistry. Frankland reported the first transmetalation reaction in 1849 while preparing ethyl zinc species from zinc metal and ethyl iodide.

It is one of the fundamental reactions of organometallic chemistry, which involves the movement of the ligand from one metal to another and has a wide range of applications in polymer chemistry and in organic reactions. It helps to produce a large number of organometallic compounds of both transition metals and main group elements. For many cross-coupling reactions, transmetalation is the rate-determining step. In a transmetalation reaction, the metal–carbon bond is activated, leading to a new metal–carbon bond. This reaction is used not only to prepare organocopper compounds but also for other organometallic compounds [10].

Lipshutz et al. reported the utilization of trimethylsilyl group as a proton surrogate for generating hexamethylsilyl stannane (Me_3Sn-$SiMe_3$), whose direct treatment with $R_2Cu(CN)Li_2$ in THF at low temperature affords higher-order trimethyl stannyl cyanocuprate [11]. Once formed, these cuprates participate in the usual cuprate coupling reactions to deliver the Me_3Sn group with virtually complete selectivity, all operations being carried out in a single flask. Hexamethylsilyl stannane is prepared from the treatment of commercially available Me_3Si-$SiMe_3$ in THF/HMPA with MeLi at low temperatures. The resulting solution of Me_3SiLi is converted to Me_3Sn-$SiMe_3$ by treatment with Me_3SnCl, and then subjected directly to the transmetalation as mentioned above.

4.3 REACTIONS WITH ORGANOCOPPER COMPOUNDS

4.3.1 SYNTHESIS OF β-METHYLTHIOBUTENOLIDES

Organocuprates are widely used in carbon–carbon bond formation reactions because of their unique reactivity toward conjugate addition. Hosomi and co-workers used the reductive capabilities of these cuprates in the synthesis of vinyl-copper compounds. In this methodology, ketene dithioacetals are reduced efficiently with the help of $Me_2Cu(CN)Li_2$ to produce vinyl copper species which on treatment with an acyl chloride, yields corresponding acylated products (Figure 4.1) [12]. Hence, this protocol offers an efficient and novel methodology for the synthesis of highly functionalized organic compounds. The reduction of these compounds leads to the generation of β-methylthiobutenolides.

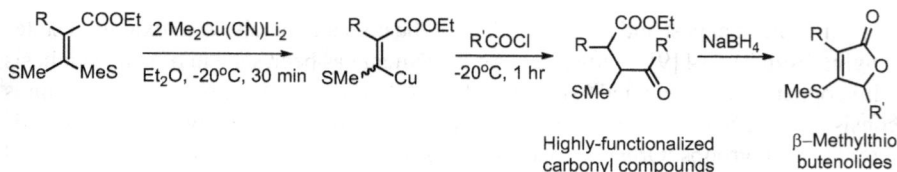

FIGURE 4.1 Synthesis of β-methylthiobutenolides.

4.3.2 EPOXIDE RING OPENING

Organocopper compounds are one of the reagents commonly used for regioselective opening of epoxide ring. Here, stereoselective synthesis of alkenes has been shown using a combination of triethylsilyloxirane and organocuprate reagent as the starting materials. Epoxide opening leads to the formation of β-silyl alcohol which on further oxidation and elimination of Et_3Si-OH group generates both the diastereomers of alkenes depending upon the condition chosen (Figure 4.2) [6].

4.3.3 CONJUGATE ADDITION

General methodology for the synthesis of catechol derivatives has been reported by Gurski and group in which 4-vinyl,2-cyclobutenone were prepared by the addition of carbon nucleophile to cyclobutendiones. This is followed by thermolysis at 100 °C which gave highly substituted aromatic compounds in good yield (Figure 4.3) [13].

The development of stereoselective carbon–silicon bond-forming reactions is highly desirable in organic chemistry. Hale et al. have reported the conjugate addition of lithium-cuprate reagent lithium bis[diphenyl(diethylamino)silyl]-cuprate $[(Ph_2(Et_2N)Si)_2CuLi]$ to unsaturated esters and ketones for the synthesis of β-carbonyl siloxanes. Stereochemical route of reaction was managed by the presence of suitable Lewis basic functionality giving the final product in excellent diastereoselectivity (Figure 4.4) [14]. The resultant products could be used as templates for stereoselective transformations that do not involve silyl transfer by accident.

Xiang and co-workers were able to synthesize the trisubstituted vinyl triflones through the addition of an organocopper reagent to acetylenic triflones [15]. Poor

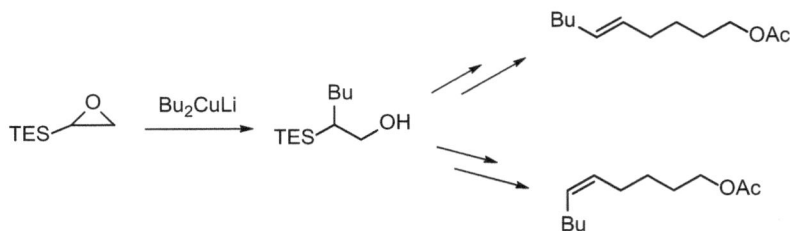

FIGURE 4.2   Epoxide ring opening using organocuprate reagent.

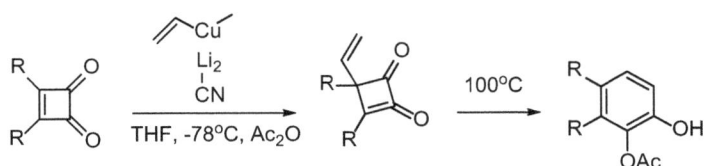

FIGURE 4.3   Addition of cuprates to symmetrically substituted cyclobutendiones.

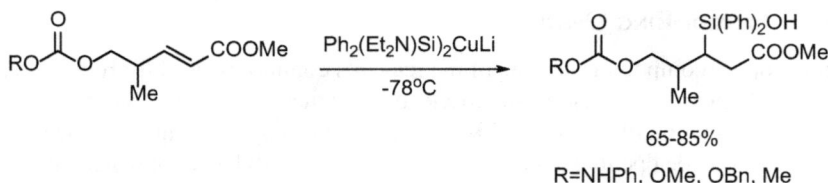

FIGURE 4.4 Conjugate addition of [Ph$_2$(Et$_2$N)Si)$_2$CuLi] to unsaturated carbonyl.

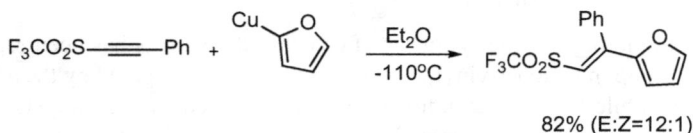

FIGURE 4.5 Addition of organocopper reagents to acetylenic triflones.

selectivity of the product (syn:anti) was seen when the corresponding lithium-cuprates were used as nucleophiles (Figure 4.5).

4.3.4 Preparation of Allenes

Allenes are the main target molecules in organic chemistry; the excellent reactivity of allenes promotes the conversion of allenes into other valuable molecules. They act as important building blocks toward the complex molecular targets and are being utilized in the synthesis of many natural and pharmaceutical products. A new methodology has been developed by Hohmann and the group for the synthesis of allenes by the 1,6-addition of organocuprates to acceptor-substituted enynes. Treatment of lithium dimethylcuprate with trimethylsilyl iodide (Lewis acid) afforded the desired allene with 45% of isolated yield (Figure 4.6) [16].

4.3.5 Coupling with Acyl Chloride

Linderman and co-workers have shown a completely selective transfer of secondary α-acetoxyalkyl groups can occur from (α-acetoxyhexyl)tricyclohexyltin but during coupling reactions, non-selective transfer of alkyl group is observed. They found that upon treatment of α-acetoxy tin with acyl chloride in presence of copper as catalyst results in the generation of the desired α-acetoxy carbonyl product in moderate yields along with few other by-products (Figure 4.7) [17].

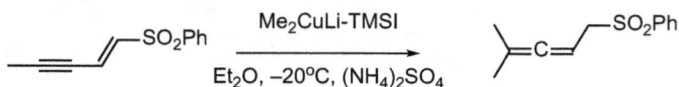

FIGURE 4.6 Synthesis of allenes.

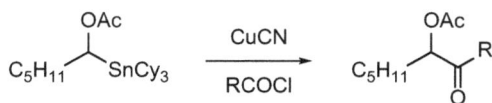

FIGURE 4.7   Coupling with acyl chloride.

4.3.6   RING-OPENING REACTION OF α,β-EPOXY SILANE

α,β-Epoxy silanes react with a wide variety of reagents *via* ring-opening α to silicon atom. These reactions have been found to have a variety of uses in organic synthesis. Hudrlik et al. have reported the use of organocopper reagent in ring-opening reaction of α,β-epoxy silane. It was observed that the treatment of epoxy silane with PhLi/CuI in 2:1 and BF$_3$.Et$_2$O afforded β-hydroxy silane in 86% yield. The further treatment of the alcohol with KH or BF$_3$.Et$_2$O gave olefin of different diastereoselectivity (Figure 4.8) [18].

4.3.7   REACTION OF ORGANOCUPRATE WITH α,β-UNSATURATED KETONE

1,4-Addition of lithium organocuprate with α,β-unsaturated ketone has turned out to be an important reaction for the formation of the new carbon–carbon bond at the β position [19]. In 1,4-addition, α carbon of α,β-unsaturated ketone forms bond with hydrogen and β carbon forms a new bond with the incoming nucleophile. It is reported that the treatment of 2,2,5-trimethyl-4-hexen-3-one with 1.5 equivalents of Me$_2$CuLi.LiI in the presence of ethoxyethane at 0 °C provided 1,4-addition product in 90% yield [20]. The effect of concentration of cuprate (Me$_2$CuLi.LiI) was tested, and it was seen that different ratios of 1,4-product were formed with high concentration and low concentration of the reagent.

4.3.8   SYNTHESIS OF β-AMINO ACID

β-Amino acids and their α-hydroxy derivatives are important because many of them occur in diverse natural products endowed with significant biological activity. Notwithstanding an apparent abundance of methods, the authors have demonstrated

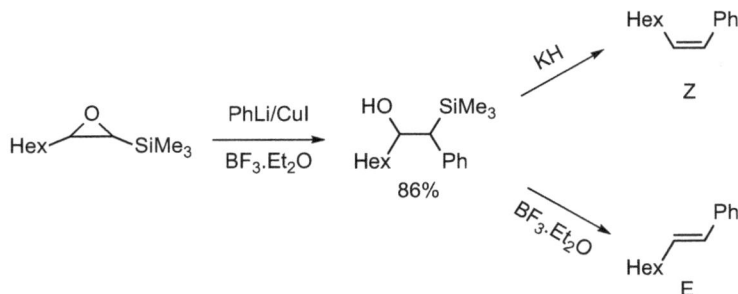

FIGURE 4.8   Ring-opening reaction of α,β-epoxy silane.

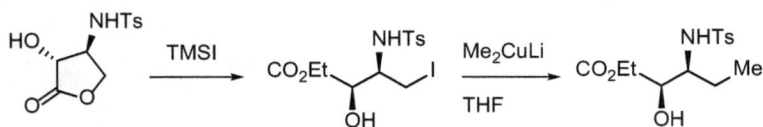

FIGURE 4.9 β-Amino acid synthesis with Gilman reagent.

the versatility of enantiomeric 3-(tosylamino)butano-4-lactones which can be obtained from L- and D-aspartic acids. The lactone serves as a versatile template for preparing amino acids and their hydroxy derivatives of 'syn' configuration in optically pure form. The key step is the introduction of alkyl chain using Gilman reagent (Figure 4.9) [21].

4.3.9 SUBSTITUTION REACTION

Linderman et al. presented a new diastereoselective method for the synthesis of α-stannyl allyl ethers. The final compound was obtained in good yield when the reaction was carried out in the presence of $BF_3.Et_2O$ with varying organocuprates at −78 °C (Figure 4.10) [22].

4.3.10 REMOTE ASYMMETRIC INDUCTION

Rakotoarisoa and group had studied the addition of $PhCu-BF_3.Et_2O$ reagent to chiral dienic acetal and observed that the regioselectivity of reaction was found to be dependent on both nature and substitution of starting acetal system (Figure 4.11) [23]. Acidic hydrolysis of the resulting enol ethers afforded the corresponding chiral δ-substituted aldehydes in moderate yields.

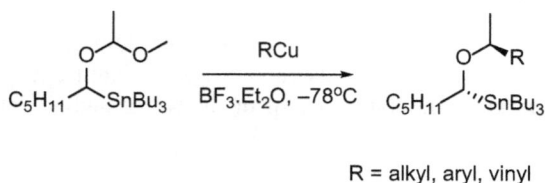

R = alkyl, aryl, vinyl

FIGURE 4.10 Substitution reaction.

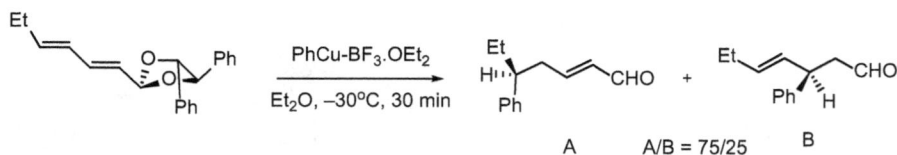

FIGURE 4.11 Synthesis of δ-substituted aldehydes using asymmetric induction.

TABLE 4.1
Synthetic Methods for Organolithium Compounds

Entry	Reaction	Reference
1.	Halogen-metal exchange	[25]
2.	Shapiro reaction	[26]
3.	*Ortho*/direct metalation	[27]
4.	Transmetalation	[10]

4.3.11 ORGANOLITHIUM COMPOUNDS

Organolithium compounds contain a bond between lithium and a carbon atom. Due to the electropositive nature of lithium, charge density is on carbon; hence it behaves as a carbanion, and the organolithium compound acts as both nucleophile and base. They are essential intermediates and effective reagents for organic transformations due to their exceptionally high reactivity. Organolithium compounds are thermally more stable as compared to other alkali metals. They are found to make coordination complexes with ethers, neutral Lewis base, amines, and alkoxides and display diverse features depending on the complexing reagent [24]. Organolithium compounds are found as oligomers in solutions, and separation occurs when they react with an electrophile. The degree of aggregation and reactivity can be controlled with the help of chelating ligands such as tetramethylene diamine and hexamethyl phosphoric triamide.

4.3.12 SYNTHETIC METHODS FOR ORGANOLITHIUM COMPOUNDS

Organolithium compounds can be prepared by various methods which are listed in Table 4.1.

4.4 REACTIONS WITH ORGANOLITHIUM COMPOUNDS

4.4.1 ADDITION REACTIONS

Parker and group have found that lithiated glycal undergoes 1,2-nucleophilic addition reaction with functionalized quinone derivative to give C-arylglycoside. The addition reaction is mediated by $BF_3.Et_2O$ and proceeds at -78 °C in THF. The reductive aromatization of intermediate is done with $Na_2S_2O_4$ in a mixture of THF/H_2O (5:2). The addition product thus obtained can be converted into papulacandin nucleus by a series of reactions including epoxidation and spiroketalization, on the other hand, hydroboration leads to the chaetiacandin framework (Figure 4.12) [28].

(R)-O-(1-Phenylbutyl)cinnamaldoxime undergoes asymmetric reaction with organolithium in presence of $BF_3.Et_2O$ to afford hydroxylamine with 95% isolated yield and 92% diastereoselectivity (Figure 4.13) [29]. The chiral hydroxylamine can be converted into N-protected amino acids via N–O bond cleavage and conversion into carboxyl functionality via a combination of Ru-NaIO$_4$. The methodology was also applied toward the preparation of ketoxime ethers and subsequent conversion to quaternary amino acids.

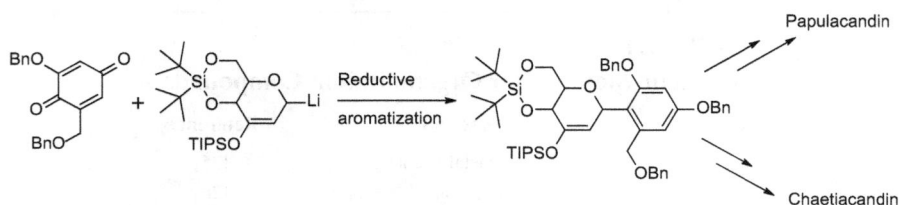

FIGURE 4.12 Addition reaction of functionalized quinone with lithiated glycal.

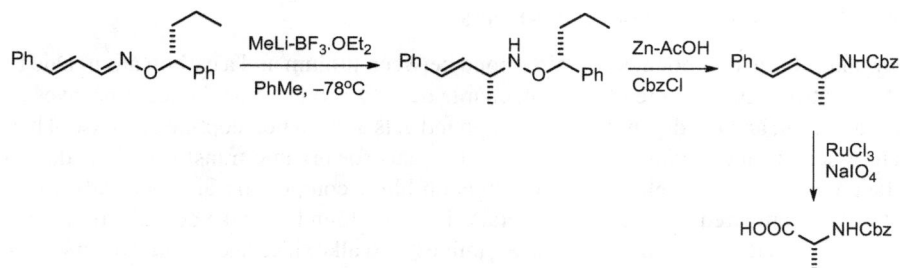

FIGURE 4.13 Synthesis of hydroxylamine with help of organolithium reagent.

Knight et al. reported the incorporation of lithiated methyl phenyl sulfone to nitrones in THF at −78 °C, followed by quenching with ammonium chloride and aqueous workup provides hydroxylamine in excellent yields and enantioselectivity. Rapid reverse-cope elimination of hydroxylamine derivative at room temperature gives pyrrolidine N-oxide as a single enantiomer (Figure 4.14) [30].

4.4.2 SYNTHESIS OF 1,4-DIKETONES

1,4-Diketones are significant organic scaffolds used as precursors for the preparation of several heterocyclic compounds. Due to this reason, many protocols have been developed for their synthesis. An effective and novel methodology was reported by Varea et al. for the synthesis of disubstituted 1,4-diketones from squaric acid derivatives. The protocol involves the addition of organolithium compound on squaric acid at room temperature followed by hydrolysis (Figure 4.15) [31].

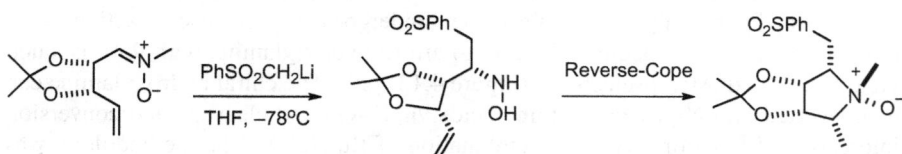

FIGURE 4.14 Reaction of nitrone derivative with lithiated methyl phenyl sulfone.

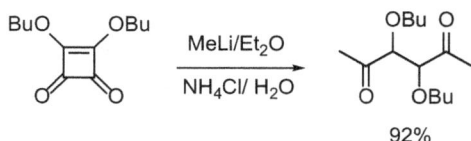

FIGURE 4.15 Synthesis of 1,4–diketones.

4.4.3 REDUCTIVE COUPLING OF ALDEHYDE TOSYLHYDRAZONES

In the growing pantheon of carbon–carbon formation reactions reductive coupling stand out because of the easy availability of corresponding starting materials. In this regard, an improved and effective method for the formation of carbon–carbon σ bonds by reductive coupling of aldehyde tosylhydrazone with organolithium reagent has been reported by Myers and the group. They found that the treatment of 1.2 equivalent alkyl lithium reagent with aldehyde tosylhydrazone at −78 °C furnished the desired adduct in excellent yields (Figure 4.16) [32].

4.4.4 PARHAM CYCLIZATION

Parham cyclization is one of the crucial methods used in the synthesis of heterocyclic compounds. Generally, it is performed by lithium–halogen exchange, followed by reaction with internal electrophile causing ring closure. Figure 4.17 shows a representative example of the application of Parham cyclization, here E are groups like CH_2Br, CH_2Cl, epoxide, etc. bearing a leaving group (Figure 4.17) [33].

4.4.5 WITTIG REARRANGEMENT

Wittig rearrangement is a type of sigmatropic rearrangement which is promoted by base. It involves the concerted reorganization of electrons during which a group attached by σ bond migrates to the terminus of an adjacent π bond. This reaction shows great stereo control and can be used early in the synthetic route to generate

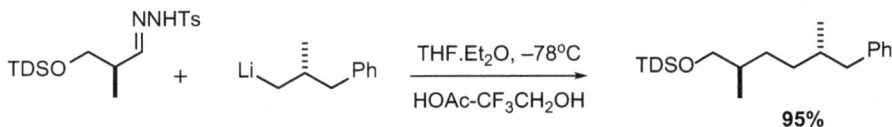

FIGURE 4.16 Reductive coupling of aldehyde tosylhydrazones.

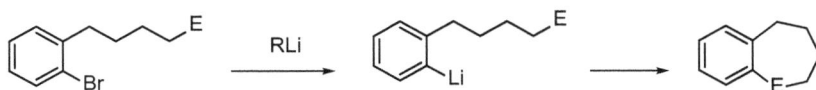

FIGURE 4.17 Parham cyclization.

stereochemistry. Rearrangement of ethers on treating with base is called Wittig rearrangement. With [2,3]-Wittig rearrangement homoallylic alcohols can be synthesized from allylic ether while [1,2]-Wittig rearrangement yield secondary or tertiary alcohols from ethers [34].

4.4.6 RAMBERG–BÄCKLAND REACTION

The Ramberg–Bäcklund reaction is a convenient method that allows a base-mediated conversion of α-halosulfones into E or Z alkenes. Z alkenes are often observed with weak bases, whereas, strong bases give predominantly E alkenes. Among the many known methods for preparing alkene Ramberg–Bäcklund reaction stands out due to the easy availability of sulfones (Figure 4.18) [35].

4.4.7 SYNTHESIS OF 1,5-DIKETONES

A new and efficient strategy for the synthesis of 1,5-diketones is reported and applied in the preparation of the fungicidal sesquiterpene (±)-α-herbertenol. In this protocol, 3,4-dihydropyranones were treated with organolithium reagent in presence of THF and good yields were achieved. The reaction was quenched with TMSCl before the hydrolytic workup and organolithium reagent which has donor group adjacent to carbon–lithium bond is functionalized (Figure 4.19) [36].

4.4.8 REARRANGEMENT REACTION

A simple and proficient method for synthesis of fused tricyclic 5–6–7 ring systems was reported by Ovaska and co-workers *via* Claisen rearrangement. The tricyclic product was obtained in 85% yield when the substrate was heated with a catalytic amount of methyl lithium (0.1 equivalent) in Ph₂O as solvent at 195 °C for 1 h (Figure 4.20) [37].

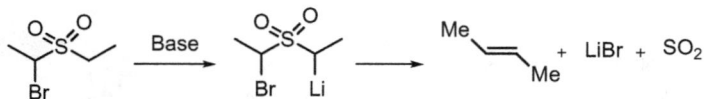

FIGURE 4.18   Ramberg Bäckland reaction.

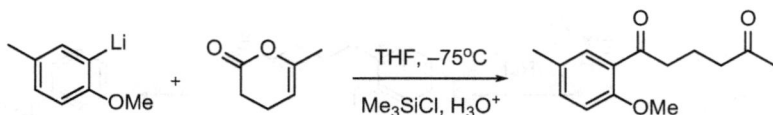

FIGURE 4.19   Synthesis of 1,5-diketones.

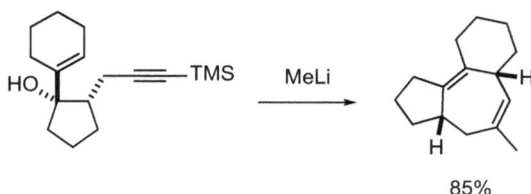

FIGURE 4.20 Rearrangement reaction with organolithium reagent.

4.4.9 SYNTHESIS OF α,β-BUTENOLIDES

Metal carbonyl-substituted compounds are useful reagents for organic synthesis. Braun et al. reported the reaction of cyclic [η^5-C$_5$H$_5$(CO)$_2$Fe] substituted enals with organolithium reagent in THF giving α,β-butenolides. The reaction proceeds *via* smooth 1,2-addition onto the aldehyde functionality in the presence of 1.05 equivalent of organolithium reagent followed by an intramolecular cyclocarbonylation (Figure 4.21) [38].

4.4.10 REACTION WITH EPOXY SILANE

Epoxy silanes are bifunctional silanes that are comprised of a reactive epoxy group and a hydrolyzable alkoxy group. Negishi et al. report a stereospecific procedure for preparing oxygenated heterocycles containing an exocyclic alkene group using Hudrlik's epoxysilane opening reactions with heteroatom nucleophiles. In this protocol epoxy silane reacts with an organolithium reagent in the presence of a base like KH or under specific reaction conditions to give alkenes in 95% yield (Figure 4.22) [39].

4.4.11 ORGANOZINC COMPOUNDS

Organozinc compounds consist of carbon–zinc bonds that are reactive compounds, hence holds the central position in organic chemistry [40]. Organozinc compounds

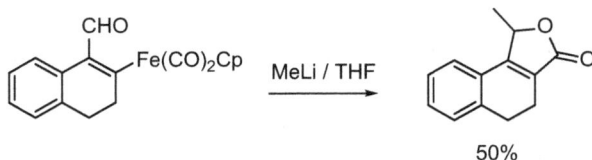

FIGURE 4.21 Synthesis of α,β-butenolides.

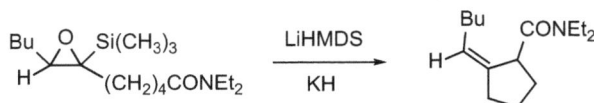

FIGURE 4.22 Reaction with epoxy silanes.

TABLE 4.2
Synthetic Methods for Organozinc Compounds

Entry	Method	Reference
1	Displacement reaction	[43]
2	Transmetalation	[44]
3	Carbocyclization	[45]
4	Oxidative addition	[44]

are readily prepared by oxidative addition of zinc to alkyl, allylic, or benzylic halides, or by transmetalation reactions. Carbon–zinc bond gets inserted to moderately polar electrophiles like aldehydes, nitriles, and ketones. Frankland discovered diethyl zinc, the first organozinc compound, in 1849 [41]. Diethyl zinc is a colorless, volatile liquid that ignites easily on exposure to air. Decomposition of organozinc compounds occurs quickly in water, so reactions should be performed under the inert atmosphere of nitrogen, carbon dioxide, or argon. Organozinc reagents can be transmetallated to other reactive organometallic species and possess good chemoselectivity. Availability of low-lying p-orbitals at the zinc center favors transmetalation with a number of transition metal complexes. There are three basic classifications of organozinc compounds: organozinc halides, organozinc, and lithium or magnesium zincates [42]. They are suitable for the preparation of polyfunctional organic molecules without using protection and deprotection steps. Hence, Organozinc compounds are not only important in organic chemistry but also work as catalysts in industrial processes.

4.4.12 SYNTHETIC METHODS FOR ORGANOZINC COMPOUNDS

The methods used for the preparation of organozinc compounds are discussed in Table 4.2.

4.5 REACTIONS WITH ORGANOZINC COMPOUNDS

4.5.1 ADDITION OF ORGANOZINC REAGENT TO β-KETO PHOSPHONATES

Wiemer et al. report that a reaction between β-keto phosphonates with allylzinc bromide results in nucleophilic addition of carbonyl group to give β-hydroxy phosphonates in excellent yield. It was observed that the addition of allylmagnesium reagent is particularly efficient in the presence of $BF_3.Et_2O$. β-Hydroxy phosphonates on further reaction with O_2 in presence of $PdCl_2$ and $Cu(OAc)_2$ furnishes oxidation of allyl group (Figure 4.23) [46].

4.5.2 SYNTHESIS OF 1,3-DIENES

1,3-Dienes have various applications, like used as chemicals in the production of rubbers and are also used as intermediates for the synthesis of many polymers,

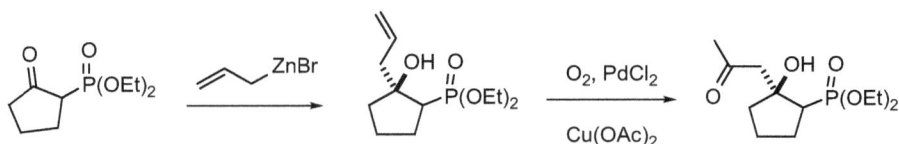

FIGURE 4.23 Addition of organozinc reagent to β-keto phosphonates.

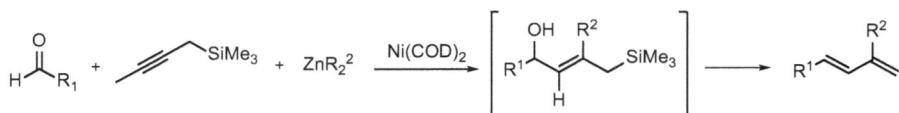

FIGURE 4.24 Synthesis of 1,3-dienes.

hence are of utmost importance. Montgomery and co-workers report the reaction between propargyl trimethylsilane with organozinc reagent to yield δ-hydroxy allyl silane. The reaction proceeds with the direct elimination of trimethylsilanol to give 1,3-disubstituted dienes in good yields (Figure 4.24) [47].

4.5.3 NEGISHI CROSS-COUPLING REACTION

Nickel catalyzed cross-coupling reaction between the organometallic reagent and organic halides is one of the main ways to prepare polyfunctional molecules, and Negishi cross-coupling reaction using alkyl zinc halides has proven to give a good yield of products. It is one of the essential methodologies for synthesizing new C–C bonds between complex synthetic intermediates. Apart from synthesizing organic molecules, the Negishi coupling reaction has various optical, magnetic, electrochemical materials, and polymer chemistry applications. Knochel et al. reported an improved nickel catalyzed cross-coupling reaction to yield *ortho*-substituted aryl ester in excellent yields. The coupled product was synthesized by the reaction between *ortho*-substituted aryl iodide and alkyl zinc iodides in the solid or solution phase [48].

4.5.4 ALKYLATIVE RING OPENING

Lautens et al. had investigated the stereochemical outcome in ring-opening reactions of oxabicyclic compounds and reported a new enantioselective transformation that significantly expands the scope and utility of this reaction. The novel method for nucleophilic ring-opening involves a reaction of oxabicyclic compound with dialkyl zinc reagent, which leads to the desired product in good yield (Figure 4.25) [49]. This method is applicable to the enantioselective synthesis of cyclohexenols, cycloheptenol, and dihydronaphthols.

4.5.5 ADDITION OF DIALKYL ZINC TO ALDEHYDES

Noyori et al. were the first to study the enantioselective catalytic addition of organozinc reagent to benzaldehyde and studied the 'ligand acceleration' term thoroughly [50].

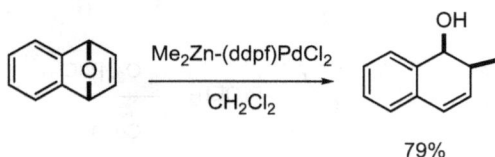

FIGURE 4.25   Alkylative ring opening.

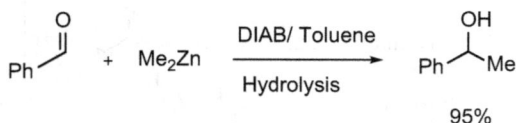

FIGURE 4.26   Addition of dialkyl zinc to aldehyde.

They found that the addition of 2 mol% of DAIB ((−)-3-exo-dimethyl amino iso-borneol) to benzaldehyde and dimethyl zinc results in the formation of secondary alcohols in 97% yield with 95% of enantioselectivity (Figure 4.26) [51].

4.5.6   SYNTHESIS OF ALLENES

Allenes are the organic compounds in which one carbon atom has double bonds with the two adjacent carbon centers. They are categorized as cumulative dienes. Varghese and the group had described the method for synthesizing substituted allenes that are further used to synthesize chiral compounds. Carbocupration of alkynyl sulfoxide, followed by homologation of vinyl copper with zinc carbenoid, lead to the formation of allyl zinc derivative that further undergoes β-elimination to form substituted allene (Figure 4.27) [52].

4.5.7   PREPARATION OF POLYFUNCTIONAL NITRILES

Organozinc halides show high functional group tolerance, hence allow easy preparation of polyfunctional organozinc compounds. Knochel et al. have observed that

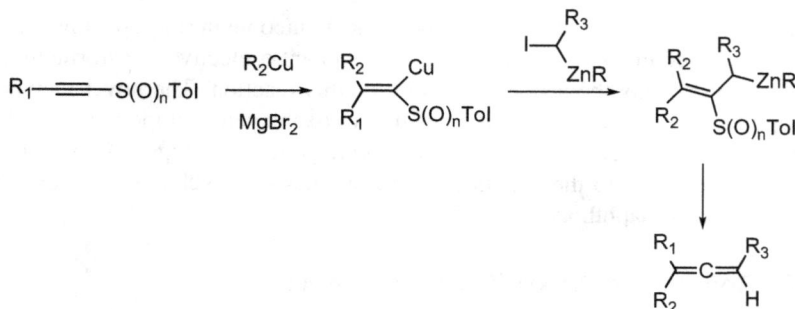

FIGURE 4.27   Synthesis of allenes.

p-toluene sulfonyl cyanide can react with various organozinc compounds to afford polyfunctional nitriles in good yield, which are further used in the synthesis of aminopyrrole, pyridazine, and pyrazolopyridazine derivatives. The reaction of *p*-toluene sulfonyl cyanide (ToSCN) with organozinc halide compound in THF solvent at room temperature yields heteroaromatic nitrile in 69% [53].

4.5.8 SYNTHESIS OF CYCLOPENTENONE

Crimmins et al. had reported the formation of cyclopentenone derivatives by the conjugate addition cyclization of zinc homoenolate to acetylenic esters under suitable reaction conditions with an isolated yield of 65% (Figure 4.28) [54]. Cyclopentenone works as excellent dienophiles in Diels–Alder reaction.

4.5.9 GEMINAL-FUNCTIONALIZATION OF CYCLOPROPANES

It was reported that the reaction of 1,1-dibromocyclopropane with n-BuLi proceeds with the formation of lithium carbenoid. Further treatment of lithium carbenoid with $ZnCl_2$ and alkyl lithium gives trans-zincate carbenoid, acid hydrolysis of which gives the cis-geminal functionalized cyclopropane derivatives as the only product (Figure 4.29) [55].

4.5.10 NUCLEOPHILIC SUBSTITUTION REACTION

The carbon–carbon bond formation is the basis of many organic reactions, and nucleophilic substitution reaction is one of the reactions for carbon–carbon bond formation. Brown et al. described the method for forming carbon–carbon bonds by a direct nucleophilic substitution reaction with a variety of carbon nucleophiles. The reaction of cyclic ether sulfones with an organozinc reagent in the presence of THF solvent and PhMgBr leads to the formation of a carbon–carbon bond at 2-position of cyclic ethers [56]. The group has tested the reaction with various solvents like THF and diethyl ether, but only THF gave the desired product. Magnesium's role in this reaction was not clear, but it was an important component as the reaction failed with

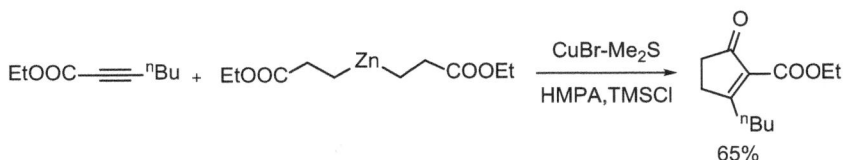

FIGURE 4.28 Synthesis of functionalized cyclopentenone.

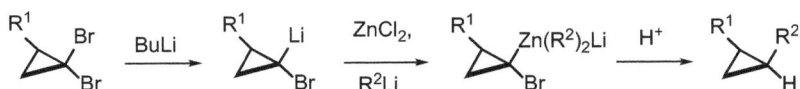

FIGURE 4.29 Gem-functionalization of cyclopropanes.

organozinc species directly prepared from organolithium species. If Grignard reagent could not be used, then magnesium bromide must be added to the reaction mixture.

4.5.11 ORGANOSELENIUM COMPOUNDS

Selenium belongs to the chalcogen family and is an essential trace element, but the relative proportion in the environment is deficient [57]. Selenium is more polarizable as compared to sulfur, so it interacts strongly with the metals. Organoselenium compounds have a carbon–selenium bond. Diethyl selenide, the first organoselenium compound, was discovered in 1836 by Lowig [58]. They provide improved oxygen resistance and corrosion inhibition efficiency as compared to Tellurium and Sulfur. Organoselenium compounds are very stable and insensitive toward the air; hence they are robust. They possess long shelf life as they do not undergo oxidation and decomposition easily [59].

Moreover, they offer the benefit of recyclability. Due to all these properties, organoselenium compounds are used as biological compounds, especially the six-membered ring organoselenium compounds and catalytic systems. Apart from this, they have a crucial role as a precursor for radicals as homolytic cleavage of carbon–selenium bond is easy which can be done either thermally or photochemically.

4.5.12 SYNTHETIC METHODS FOR ORGANOSELENIUM COMPOUNDS

Organoselenium compounds can be made by the methods mentioned in Table 4.3.

TABLE 4.3
Synthetic Methods for Organoselenium Compounds

Entry	Reactions	Reference
1		[60]
2		[61]
3		[62]
4		[63]

4.6 REACTIONS WITH ORGANOSELENIUM COMPOUNDS

4.6.1 Suzuki–Miyaura Carbon–Carbon Cross-coupling Reaction

Pd-Catalyzed Suzuki–Miyaura cross-coupling reactions have become essential tools for the construction of carbon–carbon and carbon–heteroatom bonds. In the last few decades, countless efforts have been made in the discovery and development with Suzuki–Miyaura cross-coupling chemistry and its application in pharmaceuticals and agrochemicals. The reaction occurs between aryl halide and organoborane derivatives [64]. Many organoselenium–palladium compounds have tremendous use as catalyst for this coupling reaction. Among the many catalysts known, catalyst 1 showed the best catalytic activity even with catalyst loading of only 0.04 mol% (Figure 4.30) [65].

4.6.2 Sonogashira Coupling

Another important cross-coupling reaction for carbon–carbon bond formation is the Sonogashira reaction. In this reaction aryl halides react with terminal alkynes to give sp^2–sp carbon–carbon bond [66] in the presence of a Pd-catalyst. This reaction is used in the synthesis of many natural products, pharmaceutical products, and organic materials.

Movassagh and co-workers have synthesized catalyst 2 by immobilizing an organoselenium–palladium pincer complex on polystyrene. The amount of catalyst incorporated into the polymer matrix was confirmed by ICP analysis. This catalyst was found to be very effective for Sonogashira coupling at low loading. It was also noted that less reactive and inexpensive aryl bromides and aryl chlorides have been successfully cross-coupled with terminal alkynes using low catalyst loading (Figure 4.31) [67].

4.6.3 Aldehyde–Alkyne–Amine (A³) Coupling Reaction

The A^3 reaction (also known as A^3 coupling or the aldehyde–alkyne–amine reaction), was a reaction whose name was given by Prof. Chao-Jun Li of McGill

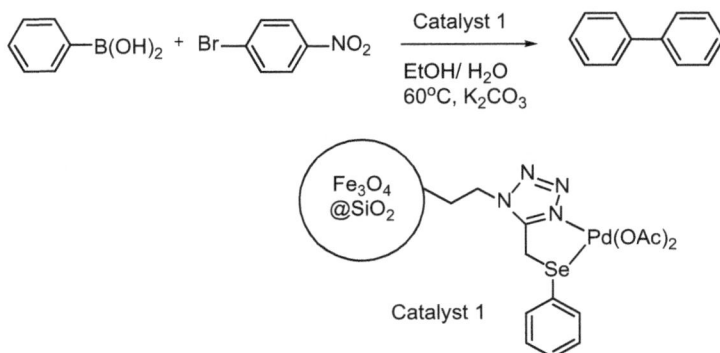

FIGURE 4.30 Suzuki–Miyaura reaction with organoselenium catalyst.

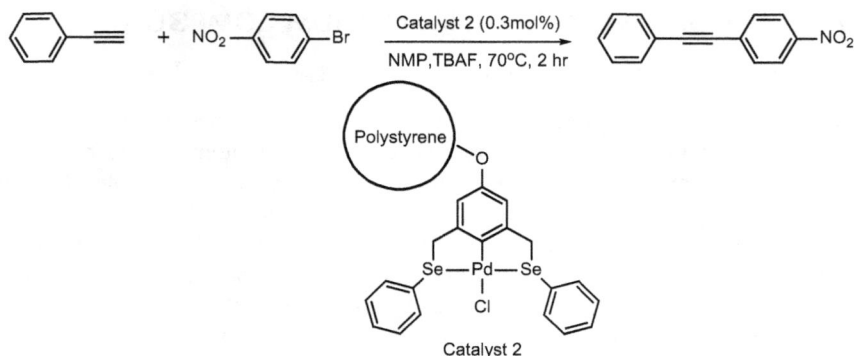

FIGURE 4.31   Sonogashira coupling.

University. It is a type of multicomponent reaction involving an aldehyde, an alkyne, and an amine which reacts to give a propargyl-amine [68]. The reaction proceeds *via* direct dehydrative condensation and requires a metal catalyst. Selenium complexes have been widely used in a variety of organic transformations. Selenium-based catalyst 3 was found to be active in catalyzing the multicomponent A^3 coupling reaction under solvent-free conditions. Hence, this reaction is eco-friendly and the organoselenium group in the catalyst makes it air insensitive (Figure 4.32).

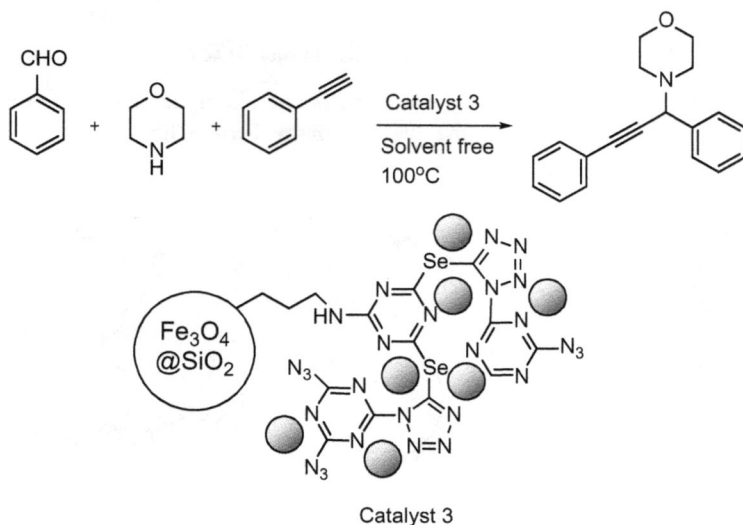

FIGURE 4.32 A^3 coupling reaction.

4.6.4 OXIDATION OF CYCLOHEXENE

Utilization of catalyst 4 in the oxidation of cyclohexene to get trans 1,2-cyclohexa-dinol was reported by Wang et al. [69]. In catalyst 4, Se is present in a hexavalent state and makes the compound air-insensitive (Figure 4.33). The hexavalent state of Se gets reduced to the divalent state during the catalytic cycle which again gets oxidized by air and comes back in hexavalent state without the utilization of hydrogen peroxide.

4.6.5 HECK REACTION

Heck reaction is one of the most well-known and most important reactions for the carbon–carbon bond formation in organic chemistry. Reaction of aryl halide with terminal alkenes gives substituted olefins in excellent yields [70, 71]. Many organoselenium ligands have been tested for the Heck reaction but catalyst 5 was found to be the most efficient for catalyzing the reaction (Figure 4.34) [72].

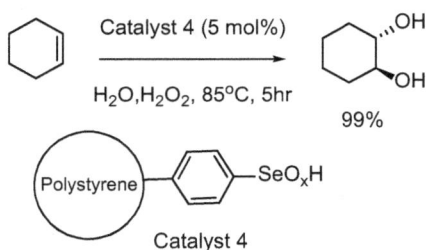

FIGURE 4.33 Oxidation of cyclohexene.

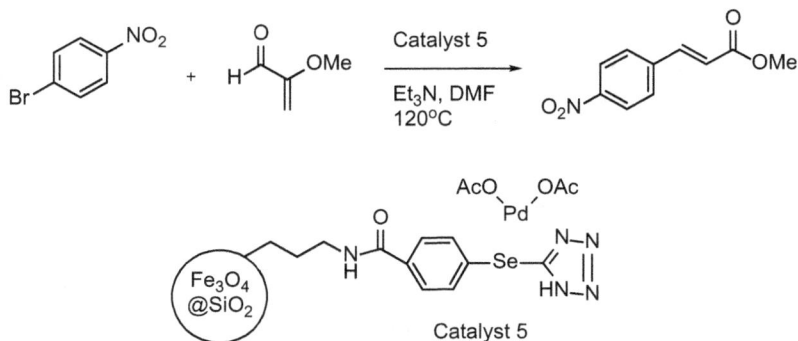

FIGURE 4.34 Heck reaction.

REFERENCES

1. Hartwig, J. F. (2010). *Organotransition Metal Chemistry*. University Science Books, Sausalito, CA.
2. Negishi, E. I. (1980). *Organometallics in Organic Synthesis*, Wiley, New York.
3. Beller, M., & Bolm, C. (1998). *Transitions Metals for Organic Synthesis*, Wiley-VCH, Weinheim.
4. Crabtree, R. H. (1988). *The Organometallic Chemistry of the Transition Metals*, Wiley, New York.
5. Kotha, S., & Meshram, M. (2018). Application of organometallics in organic synthesis. *Journal of Organometallic Chemistry*, *874*, 13–25.
6. Sudesh, K., Prachi, R., Ameta, K., & Dhrama, K. (2014). Applications of organometallic compounds as heterogeneous catalysts in organic synthesis, *Heterogeneous Catalysis*, CRC Press, London, United Kingdom 191–216.
7. Crabtree, R. H. (2005). *The organometallic chemistry of the transition metals*. John Wiley & Sons Inc., N.J. Hoboken.
8. Mudi, S. Y., Usman, M. T., & Ibrahim, S. (2015). Clinical and industrial application of organometallic compounds and complexes: A review. *American Journal of Chemistry and Applications*, *2*(6), 151–158.
9. Van Koten, G. (2012). Organocopper compounds: From elusive to isolable species, from early supramolecular chemistry with RCuI building blocks to mononuclear R2–nCuII and R3–mCuIII compounds. A personal view. *Organometallics*, *31*(22), 7634–7646.
10. (a) Lipshutz, B. H. (1994). Synthetic procedures involving organocopper reagents. In M. Schlosser (ed.), *Organometallics in Synthesis*. Wiley, Chichester; (b) Normant, J. F. (1972). Organocopper(I) compounds and organocuprates in synthesis. *Synthesis*, *1972*(2), 63–80.
11. Lipshutz, B. H., Sharma, S., & Reuter, D. C. (1990). A new transmetallation route to mixed trimethylstannylcuprates: $Me_3Sn(R)Cu(CN)Li_2$. *Tetrahedron Letters*, *31*, 7253–7261.
12. (a) Stang, P. J., Blume, T., & Zhdankin, V. V. (1993). Synthesis of enediynes by reaction of bicycloalkenyldiiodonium salts with lithium alkynyl cuprates. *Synthesis*, *1*, 35–36; (b) Hojo, M., Harada, H., & Hosomi, A. (1994). A novel and efficient generation of functionalized vinylcopper reagents and their reactions with electrophiles: Synthesis of β-methylthiobutenolides. *Chemistry Letters*, *3*, 437–440.
13. (a) Gurski, A., & Liebeskind, L. S. (1993). A new process for the regiocontrolled synthesis of substituted catechols and other 1,2-dioxygenated aromatics: Conjugate addition of vinyl-copper, aryl-copper, and heteroarylcopper reagents to cyclobutenediones followed by thermal rearrangement. *Journal of American Chemical Society*, *115*, 6101–6108; (b) Gautschi, M., Schweizer, W. B., & Seebach, D. (1994). Preparation of enantiomerically pure 4,4,4-trifluoro-3-hydroxy-butanoic acid-derivatives, branched in the 2-position or 3-position, from 6-trifluoromethyl-1,3-dioxan and dioxin-4-ones. *Chemische Berichte*, *127*, 565–579; (c) Dieter, R. K., & Alexander, C. W. (1993). Conjugate addition-reactions of α-aminoalkylcuprates prepared from organostannyl *tert*-butylcarbamates. *Synlett*, *6*, 407–409.
14. Hale, M. R., & Hoveyda, A. H. (1994). Diastereoselective heteroatom-directed conjugate addition of silylcuprate reagents to unsaturated carbonyls. A stereoselective route to carbonyl siloxanes. *Journal of Organic. Chemistry*, *59*, 4370–4374.
15. Xiang, J., & Fuchs, P. L. (1996). Alkenylation of C–H bonds via reaction with vinyl and dienyl triflones. Stereospecific synthesis of trisubstituted vinyl triflones via organocopper addition to acetylenic triflones. *Journal of American Chemical Society*, *118*, 11986–11987.

16. Hohmann, M., & Krause, N. (1995). Synthesis of allenes by 1,6-addition of organocuprates to acceptor-substituted enynes: Scope and limitations. *Chemische Berichte, 128,* 851–860.

17. Linderman, R. J., & Siedlecki, J. M. (1996). Selective copper-catalyzed coupling reactions of (α-acetoxyhexyl) tricyclohexyltin. *Journal of Organic Chemistry, 61,* 6492–6493.

18. Hudrlik, P. F., Ma, D., Bhamidipati, R. S., & Hudrlik, A. M. (1996). Ring-opening reactions of α- and β-epoxy silanes with organocopper reagents: Reaction at carbon or silicon. *Journal of Organic Chemistry, 61,* 8655–8658.

19. (a) Lipshutz, B. H., & Sengupta, S. (1992). Organocopper reagents: Substitution, conjugate addition, carbo/metallocupration, and other reactions. *Organic Reactions, 41,* 135–631; (b) Posner, G. H. (1980). *An Introduction to Synthesis Using Organocopper Reagents.* John Wiley & Sons, New York.

20. Vellekoop, A. S., & Smith, R. A. J. (1994). The mechanism of organocuprate 1,4-addition reactions with.alpha., .beta.-unsaturated ketones: Formation of cuprate-enone complexes with lithium dimethylcuprate. *Journal of the American Chemical Society, 116(7),* 2902–2913.

21. Jefford, C. W., McNulty, J., Lu, Z. H., & Wang, J. B. (1996). The enantioselective synthesis of β-amino acids, their α-hydroxy derivatives, and the *n*-terminal components of bestatin and microginin. *Helvetica Chimica Acta, 79,* 1203–1216.

22. Linderman, R. J., & Chen, S. (1995). Diastereoselective additions of copper and cuprate reagents to α-stannyl substituted mixed acetals. *Tetrahedron Letter, 36,* 7799–7802.

23. Rakotoarisoa, H., Perez, R. G., Mangeney, P., & Alexakis, A. (1996). Remote asymmetric induction. New mechanistic insights concerning the SN' and SN substitution in organocopper chemistry. *Organometallics, 15(8),* 1957–1959.

24. Mallan, J. M., & Bebb, R. L. (1969). Metalations by organolithium compounds. *Chemical Reviews, 69(5),* 693–755.

25. Leroux, F., & Schlosser, M. (2002). Der Arin-Zugang zu Biarylen mit ungewoehnlichen Substitutionsmustern. *Angewandte Chemie, 114,* 4447–4450.

26. Uyanik, M., & Ishihara, K. (2014). 6.14 Functional Group Transformations via Carbonyl Derivatives. In Gary A. Molander, & Paul Knochel (eds.), *Comprehensive Organic Synthesis II.* 573–597. Elsevier, Oxford.

27. Hessler, A., Kottsieper, K. W., Schenk, S., Tepper, M., & Stelzer, O. (2001). A novel access to tertiary and secondary *ortho*-aminophenylphosphines by protected group synthesis and palladium catalyzed P-C coupling reactions. *Zeitschrift für Naturforschung B: A Journal of Chemical Sciences, 56(4–5),* 347–353.

28. Parker, K. A., & Georges, A. T. (2000). Reductive aromatization of quinols: Synthesis of the C-arylglycoside nucleus of the papulacandins and chaetiacandin. *Organic Letters, 2,* 497–499.

29. Moody, C. J., Gallagher, P. T., Lightfoot, A. P., & Slawin, A. M. Z. (1999). Chiral oxime ethers in asymmetric synthesis. 3. Asymmetric synthesis of (*R*)-*N*-protected α-amino acids by the addition of organometallic reagents to the ROPHy oxime of cinnamaldehyde. *Journal of Organic Chemistry, 64,* 4419–4425.

30. Hanrahan, J. R., & Knight, D. W. (1998). A new strategy for the elaboration of pyrrolidine *n*-oxides using the reverse-cope elimination. *Chemical Communications, 20,* 2231–2232.

31. Varea, T., Grancha, A., & Asensio, G. (1995). A simple and efficient route to 1,4-diketones from squaric acid. *Tetrahedron, 51,* 12373–12382.

32. Myers, A. G., & Movassaghi, M. (1998). Highly efficient methodology for the reductive coupling of aldehyde tosylhydrazones with alkyllithium reagents. *Journal of the American Chemical Society, 120,* 8891–8892.

33. Parham, W. E., Jones, L. D., & Sayed, Y. (1975). Four- to seven-membered ring annulation of aryl bromides bearing *ortho* side chains having an electrophilic moiety, accomplished by halogen-metal exchange and subsequent nucleophilic ring closure. *Journal of Organic Chemistry, 40*, 2394–2399.

34. Wittig, G., & Löhmann, L. (1942). [1,2]-Wittig rearrangement. *Liebigs Annalen der Chemie, 550*, 260–268.

35. Paquette, L. A. (1968). Reaction of α-halo sulfones with strong bases to yield alkenes. *Accounts of Chemical Research, 1*, 209–216.

36. Harrowven, D. C., & Hannam, J. C. (1999). 1,5-Diketones from 3,4-dihydropyranones: An application in the synthesis of (+/−)-α-herbertenol. *Tetrahedron, 55*, 9333–9340.

37. Ovaska, T. V., Roark, J. L., Shoemaker, C. M., & Bordner, J. (1998). A convenient route to fused 5-7-6 tricyclic ring-systems. *Tetrahedron Letter, 39*, 5705–5708.

38. Möller, C., Mikulás, M., Wierschem, F., & Rück-Braun, K. (2000). Synthesis of 5-substituted α,β-butenolides by iron-promoted intramolecular cyclocarbonylation: Addition of organometallic reagents to iron-substituted enals. *Synlett, 2000(2)*, 182–184.

39. Zhang, Y., Miller, J. A., & Negishi, E. (1989). Carbon–carbon bond formation via opening of epoxysilanes with organometals containing lithium and copper. *The Journal of Organic Chemistry, 54(9)*, 2043–2044.

40. Solmi, M. V., Schmitz, M., & Leitner, W. (2019). CO_2 as a building block for the catalytic synthesis of carboxylic acids. *Horizons in Sustainable Industrial Chemistry and Catalysis, 178*, 105–124.

41. Seyferth, D. (2001). Zinc alkyls, Edward Frankland, and the beginnings of main-group organometallic chemistry. *Organometallics, 20(14)*, 2940–2955.

42. Knochel, P., & Jones, P. (1999). *Organozinc Reagents. A Practical Approach*. Oxford University Press, Oxford.

43. (a) Langer, F., Schwink, L., Devasagayaraj, A., Chavant, P. Y., & Knochel, P. (1996). Preparation of functionalized dialkylzincs via a boron-zinc exchange: Reactivity and catalytic asymmetric addition to aldehydes. *Journal of Organic Chemistry, 61*, 8229–8243; (b) Langer, F., Waas, J., & Knochel, P. (1993). Preparation and reactions of new dialkylzincs obtained by a boron-zinc transmetalation. *Tetrahedron Letter, 34*, 5261–5264.

44. Knochel, P. (1991). Organozinc, Organocadmium and Organomercury Reagents. In B. M. Trost, & I. Fleming (eds), *Comprehensive Organic Synthesis*, 211–229. Pergamon Press, Oxford

45. Meyer, C., Marek, I., Courtemanche, G., & Normant, J. F. (1993). Carbocyclization of functionalized zinc organometallics. *Synlett, 4*, 266–268.

46. Lentsch, L. M., & Wiemer, D. F. (1999). Addition of organometallic nucleophiles to β-keto phosphonates. *The Journal of Organic Chemistry, 64(14)*, 5205–5212.

47. Qi, X., & Montgomery, J. (1999). New three-component synthesis of 1,3-dienes employing Nickel catalysis. *The Journal of Organic Chemistry, 64(25)*, 9310–9313.

48. Jensen, A. E., Dohle, W., & Knochel, P. (2000). Improved Nickel-catalyzed cross-coupling reaction conditions between *ortho*-substituted aryl iodides/nonaflates and alkylzinc iodides in solution and in the solid-phase. *Tetrahedron, 56(25)*, 4197–4201.

49. Lautens, M., Renaud, J. L., & Hiebert, S. (2000). Palladium-catalyzed enantioselective alkylative ring opening. *Journal of the American Chemical Society, 122(8)*, 1804–1805.

50. Kitamura, M., Suga, S., Kawai, K., & Noyori, R. (1986). Catalytic asymmetric induction. Highly enantioselective addition of dialkylzincs to aldehydes. *Journal of the American Chemical Society, 108(19)*, 6071–6072.

51. Dimitrov, V., & Nacheva, K. M. (2009). Enantioselective organozinc-catalysed additions to carbonyl compounds: Recent developments (review). *Journal of the University of Chemical Technology and Metallurgy, 44(4)*, 317–332.

52. Varghese, J. P., Knochel, P., & Marek, I. (2000). New allene synthesis via carbocupration–zinc carbenoid homologation and β-elimination sequence. *Organic Letters*, 2(*18*), 2849–2852.
53. Klement, I., Lennick, K., Tucker, C. E., & Knochel, P. (1993). Preparation of polyfunctional nitriles by the cyanation of functionalized organozinc halides with *p*-toluenesulfonyl cyanide. *Tetrahedron Letters*, *34*, 4623–4626.
54. Crimmins, M. T., Nantermet, P. G., Trotter, B. W., Vallin, I. M., Watson, P. S., McKerlie, L. A., Reinhold, T. L., Cheung, A. W. H., Dedopolou, D., Gray, J. L. & Stetson, K. A. (1993). Addition of zinc homoenolates to acetylenic esters and amides: A formal [3 + 2] cycloaddition. *The Journal of Organic Chemistry*, 58(*5*), 1038–1047.
55. Harada, T., Katsuhira, T., Hattori, K., & Oku, A. (1993). Stereoselective carbon–carbon bond-forming reaction of 1,1-dibromocyclopropanes via 1-halocyclopropylzincates. *The Journal of Organic Chemistry*, 58(*11*), 2958–2965.
56. Brown, D. S., Bruno, M., Davenport, R. J., & Ley, S. V. (1989). Substitution reactions of 2-benzenesulfonyl cyclic ethers with carbon nucleophiles. *Tetrahedron*, *45*, 4293–4308.
57. Řezanka, T., & Sigler, K. (2008). Biologically active compounds of semi-metals. *Studies in Natural Products Chemistry*, *35*, 835–921.
58. Löwig, C. J. (1836). About hydrogen sulfide and selenium hydrogen ether. *Annalen der Physik*, *37*, 550–553.
59. Gorup, L. F., Perlatti, B., Kuznetsov, A., Nascente, P. A. D. P., Wendler, E. P., Santos, A. A. D., Barros, W. R. P., Sequinel, T., Tomitao, I. D. M., Kubo, A. M., Longo, E., & Camargo, E. R. (2020). *RSC Advance*, *10*, 6259.
60. (a) Zade, S. S., Singh, H. B., & Butcher, R. J. (2004). The isolation and crystal structure of a cyclic selenenate ester derived from bis(2,6-diformyl-4-tert-butylphenyl)diselenide and its glutathione peroxidase-like activity. *Angewandte Chemie International Edition*, 43(*34*), 4513–4515; (b) Zade, S. S., Panda, S., Singh, H. B., Sunoj, R. B., & Butcher, R. J. (2005). Intramolecular interactions between chalcogen atoms: Organoseleniums derived from 1-bromo-4-tert-butyl-2,6-di(formyl)benzene. *The Journal of Organic Chemistry*, 70(*9*), 3693–3704.
61. Iwaoka, M., & Tomoda, S. (1992). Catalytic conversion of alkenes into allylic ethers and esters using diselenides having internal tertiary amines. *Journal of the Chemical Society, Chemical Communications*, (*17*), 1165.
62. (a) Panda, A., Mugesh, G., Singh, H. B., & Butcher, R. J. (1999). Synthesis, structure, and reactivity of organochalcogen (Se, Te) compounds derived from 1-(N,N-dimethylamino)naphthalene and N,N-dimethylbenzylamine. *Organometallics*, 18(*10*), 1986–1993; (b) Mugesh, G., Panda, A., Singh, H. B., Punekar, N. S., & Butcher, R. J. (1998). Diferrocenyl diselenides: Excellent thiol peroxidase-like antioxidants. *Chemical Communications*, (*20*), 2227–2228.
63. Besev, M., & Engman, L. (2002). Diastereocontrol by a hydroxyl auxiliary in the synthesis of pyrrolidines via radical cyclization. *Organic Letters*, 4(*18*), 3023–3025.
64. Devendar, P., Qu, R.-Y., Kang, W.-M., He, B., & Yang, G.-F. (2018). Palladium-catalyzed cross-coupling reactions: A powerful tool for the synthesis of agrochemicals. *Journal of Agricultural and Food Chemistry*, *66*, 8914–8934.
65. Rangraz, Y., Nemati, F., & Elhampour, A. (2019). A novel magnetically recoverable palladium nanocatalyst containing organoselenium ligand for the synthesis of biaryls via Suzuki-Miyaura coupling reaction. *Journal of Physics and Chemistry of Solids*, *138*, 109251.
66. Karak, M., Barbosa, L. C. A., & Hargaden, G. C. (2014). Recent mechanistic developments and next generation catalysts for the Sonogashira coupling reaction. *RSC Advances*, 4(*96*), 53442–53466.

67. Mohammadi, E., & Movassagh, B. (2018). A polystyrene supported [PdCl–(SeCSe)] complex: A novel, reusable and robust heterogeneous catalyst for the Sonogashira synthesis of 1,2-disubstituted alkynes and 1,3-enynes. *New Journal of Chemistry, 42(14)*, 11471–11479.

68. Jesin, I., & Nandi, G. C. (2019). Recent advances in the A3 coupling reactions and their applications. *European Journal of Organic Chemistry, 2019(16)*, 2704–2720.

69. Wang, Y., Yu, L., Zhu, B., & Yu, L. (2016). Design and preparation of a polymer resin-supported organoselenium catalyst with industrial potential. *Journal of Materials Chemistry A, 4(28)*, 10828–10833.

70. Jagtap, S. (2017). Heck reaction—State of the art. *Catalysts, 7(9)*, 267.

71. Ding, Y. H., Fan, H. X., Long, J., Zhan, Q., & Chen, Y. (2013). The application of Heck reaction in the synthesis of guaianolide sesquiterpene lactones derivatives selectively inhibiting resistant acute leukemic cells. *Bio organic and Medicinal Chemistry Letters, 23*, 6087–6092.

72. Nemati, F., Rangraz, Y., & Elhampour, A. (2018). Organoselenium-palladium(II) complex immobilized on functionalized magnetic nanoparticles as a promising retrievable nanocatalyst for the "phosphine-free" Heck-Mizoroki coupling reaction. *New Journal of Chemistry, 42*, 15361–15371.

5 Solid-supported Catalyst in Heterogeneous Catalysis

Garima Sachdeva, Dipti Vaya, and Varun Rawat
Amity University Haryana, Gurugram (Haryana), India

Pooja Rawat
Kyung Hee University, Yong-In, Gyong-gi, Republic of Korea

CONTENTS

DOI: 10.1201/9781003126270-5

5.1   INTRODUCTION

Heterogeneous catalysis is an integral part of chemical industries and modern energy, through which a gas or a liquid phase reaction is carried out over a solid catalyst. Solid-supported catalysts play an immense role in the large-scale processes for converting fuels, pollutants, and chemicals [1–5]. Although there are many advantages of using a solid-supported catalyst, some disadvantages are limited design improvements, poor accessibility of the organic substrates to the active sites, and the use of severe reaction conditions such as high temperatures and pressures. Homogeneous catalysts, on the other hand, are impervious to these drawbacks. Hence, it is advantageous to combine the advantages offered by homogeneous catalysts, namely high activity and selectivity, with the ease of separation of heterogeneous catalysts. One approach to this is to anchor homogeneous catalysts to inert support so that the ligand sphere of the metal complex is unchanged. The components of the hybrid catalytic systems must be tailored to meet reaction requirements. Supported catalysts must exhibit high selectivities and reasonable turnover numbers. The catalyst support must be mechanically strong and stable at elevated temperatures. Above all, the support must be chemically inert to the reactants, products, and solvents employed in the reaction. The active site in supported catalysts must be stable under the reaction conditions and soluble in the reaction medium.

Solid catalysts consist of many elements and phases, which can be either crystalline or amorphous. They are categorized into main components, promoters, and supports (carriers). Main components take part in the catalytic transformation of molecules, promoters alter and enhance the performance of principal components. At the same time, supports are spread evenly on the surface of principal components and promoters, responsible for increasing the surface area, hence providing chemical and mechanical stability to the catalysts.

Solid materials, including metals, metal oxides, and metal sulfides, are classified as catalysts. In industries, few catalytic materials used are simple such as pure metals and binary oxides. Heterogeneous catalysts activity can be made better with the help of supports. Catalyst support plays a vital role in bringing out the catalytically active center, contributing to industrial processes. The majority of heterogeneous solid catalysts are base or basic oxides coated over a large surface area. Generally, solid-base catalysts are found to be more active than solid-acid catalysts. Various classifications of solid catalysts are unsupported catalysts (bulk), supported catalysts, coated catalysts, confined catalysts, polymerization catalysts, etc. but our primary focus will be on solid-supported catalysts. Solid catalysts can be separated easily from the reaction mixture and reused, making them ideal for catalysis.

In many industrial processes, supported catalysts have a crucial role. New techniques are being carried out for the synthesis of the supported catalyst. By heterogenization methods, supporting materials can either be obtained with transition metals or Schiff's base ligands [6]. Two broad types of supports have been used to anchor transition metal complexes, organic polymers, and inorganic oxides. Commonly used polymeric supports include polystyrene, polypropylene, polyacrylates, and polyvinyl chlorides. Nowadays, silica, carbon, clay, zeolites, metal oxide polymers, and other mesoporous materials are being utilized as inorganic solid supports [7–8].

Polymers can be prepared with a wide range of physical properties. As a result, their porosity, surface area, and solution characteristics can be altered by varying the degree of crosslinking case of polystyrene, variation in the degree of crosslinking allows for almost continual change from a virtually soluble material at 2% crosslinking to a completely insoluble material at 20% crosslinking. This allows different selectivity through control of diffusion of reactants within the polymer. The principal disadvantages of polymers are their poor heat transfer ability and, in many cases, their poor mechanical properties, which prevent them to be used in stirred reactors. Also, a decrease in diffusion properties generally results in a decrease in catalyst activity. Significant advantages of inorganic supports are their better mechanical and thermal stabilities coupled with good heat transfer properties. Most inorganic supports are more robust to high pressures and more resilient to changes in solvent polarity. Inorganic supports also exhibit inertness to reactants and products. Polymer-supported catalysts can also deactivate through intermolecular aggregation or condensation or by chelation of the metal. This situation can also arise in a highly crosslinked polymer due to the short-range flexibility of the polymer. Rigid inorganic matrices circumvent these deactivation processes. The swelling of polymers under variable temperature and solution conditions can make the practical control of diffusion variables difficult. This is not the case for inorganic substrates.

Mesoporous supports are found to be ideal catalyst support due to the 3-D open pore network structure, reusability, high surface area, and porosity, and their interconnected and uniformly arranged pores offer good and interactive surfaces between catalysts and reactants [9–17]. The shape of the solid catalysts is designed based on mechanical strength, thermal strength, mass, and heat transfer. Supported catalyst offers high surface area and stabilizes the dispersion of active components. It is often considered that supports are inert in nature, but they might actively interfere with the catalytic process. Highly dispersed noble metals supported on the surface of the acidic carrier is an example of active interaction between support and active phase.

Some examples of support materials are listed in Table 5.1 along with their characteristics, advantages, and disadvantages [18].

5.2   SUPPORTED CATALYSTS

Catalytically active substances are scattered over the high surface area of the support material. The foremost characteristic of supported catalyst is that the minor part is formed by the active material and is deposited on the support surface. In some instances, support is less or more inert like in α-alumina, ceramics, and porous glass, while in some cases, support participates in catalytic reactions as in the bifunctional catalytical systems [19–21]. Furthermore, several supports can bring about changes in catalytic properties of the active phase, which can decrease the strong metal–support interaction. The heterogeneous catalysts with supports like aluminum oxide,

TABLE 5.1

Characteristics, Merits, and Demerits of Different Catalytic Supports

Support	Characteristics	Merits	Demerits
Alumina	• Hardness • High melting point and compression strength • Chemical and abrasion-resistant • High thermal conductivity	• Narrow pore size • Thermally stable • High surface area and pore volume	• Hydrolysis rate of alumina precursor is difficult to control
Zeolite	• Microporous • Good electron conductivity • Inertness • Bifunctional catalyst	• Effective • Less corrosion • High thermal stability • No waste disposal problem	• Difficult to use shape selectivity • Irreversible adsorption of heavy secondary products
Silica	• Hardness • Affinity in forming large complex • Found in nature and living organisms	• Highly stable • High selectivity and efficiency • Good mechanical strength	• Low compatibility • Chances to form agglomerates
Carbon	• Non-metallic • Porous structure • Tetravalent	• Large surface area • Good elasticity • High mechanical strength and thermal stability • Inertness	• Expensive • Release of greenhouse gases during pyrolysis

titanium oxide, zinc oxide, etc., are used because of the availability and low cost for the synthesis. The active phase (oxide, metal, sulfide) undergoes support and active phase interactions, which can be determined by surface free energies and by the interfacial energy between the active phase and support. For example, the chemisorption capacity of Pt-TiO_2 (supported metals) can inhibit the reaction of supported metal oxides like Nickel silicates, Nickel aluminates, etc. [19–20].

5.3 SOLID SUPPORTS

Presently, several supports are available, having a wide range of surface area, porosity, shape, and size. In order to acquire high surface area and stabilize highly dispersed active phase, supports are often porous materials possessing high thermostability. Usually used supports are binary oxides which include SiO_2, TiO_2, ZrO_2 (tetragonal), MgO, etc., and ternary oxides like Zeolites, SiO_2-Al_2O_3. Other probable catalyst supports are aluminophosphates, calcium aluminate, kieselguhr, etc. Different variants of carbon such as activated carbon, charcoal can also be used as supports unless oxygen is needed in feed at high temperatures.

5.3.1 SILICON CARBIDE SUPPORT

Silicon carbide (SiC) is an important example of catalyst support due to its high thermal stability and mechanical strength [22]. It is prepared with a porous structure and high surface area through templating [23]. SiC shows low density, extreme specific strength, and outstanding temperature stability. Hence, it is the best support that can be used during the high-temperature catalysis process. It can be altered for particular catalytic applications by the addition of metals. SiC combines the properties of both oxide and carbon-based support. The chemical and physical properties of SiC support make it the best material to be utilized for applications in the field of catalysis when compared with classical materials.

5.3.2 METAL OXIDE SUPPORTS

Metal oxide supports consist of at least one metal oxide component dispersed on the oxide support surface [24, 25]. Generally, active oxides are transition metal oxides, while support oxides are γ-alumina, SiO_2, TiO_2, ZrO_2, and carbon. V_2O_5/TiO_2 works as the active catalyst for the oxidation of o-xylene to phthalic anhydride [26, 27]. V_2O_5-MoO_3-TiO_2 and V_2O_5-WO_3-TiO_2 are used for the selective reduction of NOx emission with NH_3 in tail gas [28, 29]. Supported vanadium catalysts can adapt themselves for oxidation reactions. Re_2O_7/Al_2O_3 is an effective metathesis catalyst [30]. For the alkane dehydrogenation and dehydrocyclization, a combination of Cr_2O_3-Al_2O_3 and ZrO_2-Al_2O_3 has been used extensively [31]. All the transition metal oxides possess low surface free energies than the typical support materials [24, 32]. Thus, they are likely to spread over the support surface and frame highly dispersed active oxide over layers. Supported oxide catalysts are called monolayer catalysts, though the support surface is incompletely covered, even the loadings equal to or greater than the theoretical monolayer coverage is good enough for catalysis. The possible reason for this is that most of the transition metal oxides form 3-D islands on the surface of support which possess the same structure as molecular polyoxo compounds [24, 25].

5.3.3 SURFACE-MODIFIED OXIDES AS SUPPORTS

Surface properties of oxides (acidity and basicity) can be modified by depositing the modifiers. The acidic strength of alumina is dramatically increased either by incorporating chloride ions into or on the surface or during impregnation with the solutions having chloride precursors of active components [33]. Chlorinated alumina is also attained by the surface reaction with carbon tetrachloride [34]. Powerful basic materials support alkali metal compounds on the alumina surface [35]. Sulfation of tetragonal ZrO_2 gives strong solid acids that have super acidic properties [36, 37] as they also catalyze the isomerization of n-alkanes to iso-alkanes at low temperatures as carried out by tungstate ZrO_2.

5.3.4 SULFIDE SUPPORTS

In petroleum refining industries, sulfide catalysts are generally used for hydroprocessing applications such as hydrosulfurization and hydrodenitrogenation. Usually, sulfide catalysts used in industries are obtained from oxides of group VIB (Mo or W), and group VIII (Co or Ni) supported on γ-alumina [38]. De Beer *et al.* discovered that there is no need for selective use of alumina support in hydrosulfurization [39, 40]. Even carbon also exhibits good properties as support for molybdenum and tungsten sulfide catalysts.

Co and Ni sulfides have shown hydrodesulfurization activity when supported on carbon which is basically higher if measured for Mo or W. Hence, Co (Ni) present in Co(Ni)-Mo(W) sulfide catalyst not only act as a promoter for MoS_2 or WS_2 but also as active phase.

5.3.5 METAL SUPPORTS

Metals possess high surface energy [32] and have the susceptibility to reduce the surface areas by partial growth. For the employment as a catalyst, they are usually dispersed on the high surface area supports, especially oxides like alumina, with the purpose to stabilize small nanosized particles under the suitable reaction condition [33, 41]. Bimetallic supported catalysts have two different metals that can either be miscible or immiscible as macroscopic bulk alloys. In industries, supported metal catalysts are usually utilized as macroscopic spheres or cylindrical extrudates.

Specific applications of Pt, Pd, Ni, Fe, and Co supported on Al_2O_3, active carbon, or SiO_2 include hydrogenation and dehydrogenation reactions. Silver on Al_2O_3 is used on ethene epoxidation. Pt supported on chlorinated alumina is a bifunctional catalyst employed in catalytic refining and isomerization of petroleum fraction. Modifying the supported Pt catalyst with cinchona additives is used for enantioselective hydrogenation of α-keto ester [42].

5.4 HYBRID CATALYSTS

Hybrid catalysts merge the homogeneous and heterogeneous catalytic transformations. The purpose is to join the positive features of homogeneous catalysts or enzymes concerning activity, selectivity, and variability in electronic and steric properties by the appropriate type of ligand (involving chiral ligand) [43] with heterogeneous catalysts having properties like ease of separation and catalyst recovery. This could be obtained by heterogenization or immobilizing active metal complexes, organometallic compounds, or enzymes on the solid support.

Dendrimers that are effective at the end of dendritic arms can be utilized for the immobilization of metal complexes [44]. Immobilized homogeneous catalysts are valid for selective oxidation and hydrogenation reactions, and they are also effective in asymmetric synthesis [45].

5.5 TYPES OF HYBRID CATALYSTS

5.5.1 HETEROGENEOUS–HOMOGENEOUS HYBRID

Heterogeneous–homogeneous hybrid catalysts are also known as immobilized molecular catalysts [46]. Benefits of using such catalysts are their eco-friendly chemical production because of the long-term stability of such catalysts in storage and operating conditions, easy product separation, low waste and contamination, and their reusability.

The main challenge in these types of catalysts is the conservation of internal activity and selectivity of the homogeneous component during immobilization which can alter the chemical and electronic structure of the catalytic center. Another challenge is to prevent the leaching of the active component.

5.5.2 Heterogeneous–Enzyme Hybrid

These catalysts are applied in the industrial production of chemicals [47]. Usually, these hybrids are achieved by support-bound enzymes, entrapment, and support-free cross-linking [47a, 48].

Support-bound enzymes comprise enzymes adsorbed onto the support or are covalently attached to the support. The rate of adsorption of enzymes on the solid support is spontaneous due to the Van der Waals interactions, ionic and chelating interactions with the support. Even the enzymes can be immobilized by trapping them in a matrix or a semipermeable membrane. The extraordinary and distinctive properties of enzymes allow for the different methodology of enzyme immobilization, i.e., support-free cross-linking, which gives highly concentrated enzyme activity, excellent stability, and low production costs by preventing the additional support.

5.5.3 Ship-in-a-bottle Catalyst

Ship-in-a-bottle catalyst, also known as tea-bag catalyst, is a metal complex trapped physically in the restricted spaces of zeolite cages. The trapped complexes are supposed to retain their numerous solution properties. Catalytic efficiency can be altered in a synergic manner by shape selectivity, electrostatic environment, and acid-base properties of the zeolite host. In zeolite cages, ligands for metal complexes are dimethyl glyoxime, ethylenediamine and different Schiff's bases, and porphyrins [49]. Zeolite encapsulated complexes are used as model compounds for mimicking the enzymes.

There are three routes to achieve the entrapped complexes:

I. Reaction of a transition metal with a preformed flexible ligand inserted into zeolite cages.
II. Collecting the ligand from smaller species inside the cavities of zeolite.
III. Synthesis of zeolite structure around the preformed transition metal complex.

5.5.4 Polymerization Catalyst

Polymerization catalysts are routinely used to regulate the incorporation of one or more monomers into a polymer chain, which can, in turn, dictate the mechanical properties of the resultant material. For instance, Ziegler–Natta catalysts are combined solid and liquid compounds with transition metals like titanium and vanadium [50]. Titanium tetrachloride with alkyl aluminum compounds is functional for olefin polymerization. Powerful catalysts are generated by supporting $TiCl_4$ on solids like

$MgCl_2$, SiO_2, or Al_2O_3 to enhance the amount of active titanium. Nowadays, the production of Ziegler–Natta catalysts is carried out by ball milling with about 5% $TiCl_4$, and the co-catalyst is triethylaluminium. Phillips catalyst comprises of hexavalent surface chromate on a silica support having high surface area. Ethylene and other hydrocarbons help in the reduction of Cr^{+6} to Cr^{+2} or Cr^{+3}.

Single-site catalysts utilizing metallocene as active species are established. The activity of these materials is increased dramatically by doing the activation with methylaluminoxane. The catalytic efficiency is more than that of Ziegler–Natta catalysts and Phillips catalysts.

5.5.5   COATED CATALYSTS

One cause of deactivation in heterogeneous catalysts is that the metallic nanoparticles can aggregate over time, leading to a loss in catalytic activity and selectivity. In particular, palladium catalysts (Pd) are known to aggregate easily and form Pd black. As a result, much effort has been put into finding new catalytic systems, which effectively combine the advantages of both heterogeneous and homogeneous catalysts. Besides bulk and supported catalysts, coated catalysts are the third class of catalysts. In comparison to conventional catalyst geometries like powders, tablets, spheres, and rings, the coated catalysts are catalytically efficient and an active layer is used on an inert structured surface. The utilization of coated catalysts is becoming very popular nowadays. The active layer comprises of bulk or supported catalyst.

The examples of such systems are mentioned below [51–53]:

- Structured packings
- Fibers and cloths
- Eggshell catalysts deposited on inert carrier
- Foams and sponges
- Catalytic wall reactors

The benefits of using coated catalysts are – high specificity at low diffusion length, most efficient mass transfer from fluid phase to solid catalysts layer, optimal use of active mass, and low-pressure drop.

5.6   SOME EXAMPLES OF SUPPORTED CATALYSTS

As the properties of the supported catalyst are already discussed, here we will be dealing with the most important and common examples of supported catalysts that are utilized in many reactions.

5.6.1   TITANIUM OXIDE AS A CATALYST SUPPORT IN HETEROGENEOUS CATALYSIS

Most of the heterogeneous catalysts are unstable, as during the reaction, the particles agglomerate and block the catalyst's active site. To overcome this challenge, TiO_2 as the support for the heterogeneous catalyst is established due to its high surface area, which stabilizes the catalyst in the mesoporous form.

The properties of TiO_2 like non-toxicity, high effectiveness, extended photostability, good mechanical resistance, and stability in acidic and oxidative environments make it essential support in heterogeneous catalysis. TiO_2 enhances the catalytical performance; therefore, it can be employed for dehydrogenation, hydrodesulfurization, water gas shift, and thermal catalytic performance [54, 55]. Among distinct material candidates, TiO_2-based catalyst support materials possess excellent properties [56a]. The TiO_2 catalyst support possesses excellent resistance toward corrosion in different electrolytic media.

Metal nanoparticle on titanium oxide support study is essential in heterogeneous catalysis because of the size and interactive nature of metal nanoparticles with TiO_2 support [56b]. The interaction affects the catalytic activity and selectivity of a heterogeneous metal catalyst. Among the changes done in TiO_2, anatase is the most commonly used catalyst support because of its high specific surface area and intense interaction with metal nanoparticles [55a, 57]. The rutile form is not significant, and only a few results are reported in the literature with rutile support.

5.6.2 Au/TiO_2 Heterogeneous Catalyst

Gold is an excellent catalyst for alcohol oxidation by molecular oxygen in the liquid phase having high activity, selectivity, and resistance toward deactivation [58]. The catalytic efficiency of Au/TiO_2 is decided by the partial size and support properties [58b]. TiO_2 is an excellent support for Au heterogeneous catalyst system because of its intense interaction with the metal, chemical support, and acid–base properties [56a, 59].

Au/TiO_2 support system is used to oxidize primary alcohols into carboxylic acids [60] and is also used to remove CO and NOx. Major factors responsible for the catalytic performance of Au/TiO_2 are porosity and phase transformation. It is considered that different TiO_2 crystalline phases could influence the interaction of support-metal, Au-size, oxidation state, and Au dispersion in the heterogeneous catalytic system. To show the higher performance of the catalyst system, Au particles are to be present in a cluster size of less than 5 nm.

5.6.3 Ni/TiO_2 Heterogeneous Catalyst

Ni-Supported heterogeneous catalyst is one of the main Ni-based catalysts because of the strong interaction between the catalyst support TiO_2 and Ni metal [61]. This catalyst shows the best activity toward the acetophenone hydrogenation due to strong interaction between the Nickel and rutile surface over the Ni/TiO_2 [62]. It was observed that the hydrogenation process of maleic anhydride is highly affected by the calcination temperature of Ni/TiO_2.

5.6.4 TiO_2 Support in Bimetallic Heterogeneous Catalysis

The easy availability, affordability, and non-toxicity of TiO_2 as a powerful solid with photochemical stability make it an excellent support for heterogeneous bimetallic catalysis. TiO_2 can interact with the bimetallic system through the formation of

Ti^{3+} charge. Even the electrical conductivity of TiO_2 is because of the Ti^{3+} ions. An example of a bimetallic system having TiO_2 support is $CoMn/TiO_2$ which is used for the Fischer–Tropsch synthesis to produce C2–C4 olefins [63]. $Au-Cu/TiO_2$ was studied for the total oxidation of methane, ethane, and epoxidation of propane. Apart from the applications mentioned above, TiO_2 supports are also utilized in photocatalysis, small molecules transformation, and organic reactions, and are also used as electrodes.

5.6.5 HETEROGENEOUS METAL CATALYSTS FOR OXIDATION REACTIONS

In chemical industries, oxidation reactions play a vital role in the synthesis of specific compounds. Heterogeneous catalysts enhance the oxidation reactions either by TBHP (tert-butyl hydroperoxide) or hydrogen peroxide [64]. Here, we will be discussing some selected oxidation reactions catalyzed by the supported metal catalysts.

5.6.6 CONVERSION OF GLUCOSE TO GLUCONIC ACID

Oxidation reactions play a crucial role in chemical industries to produce many organic compounds. Homogeneous catalysts have been widely utilized for the oxidative process in the manufacturing of bulk and fine chemicals, but due to certain limitations, heterogeneous catalysts are being used in place of homogeneous catalysts to obtain high yield and fewer by-products.

Aerobic oxidation of glucose to gluconic acid (Figure 5.1) has achieved attention due to the water-soluble cleansing properties and its use in food additives. Earlier, glucose oxidation was carried out using biochemical pathways that were multistep procedures, expensive, and time-consuming; hence the growth of catalytic route is another way for the large-scale production of gluconic acid. Catalysts that promote the oxidation of glucose to gluconic acid are listed below in Table 5.2.

5.6.7 OXIDATION OF CARBON MONOXIDE

Carbon monoxide (CO) has now grown up as an important research area as it is involved in several procedures like the synthesis of methanol, water gas shift, CO_2 lasers, and automotive exhaust controls. CO gas is harmful to the environment

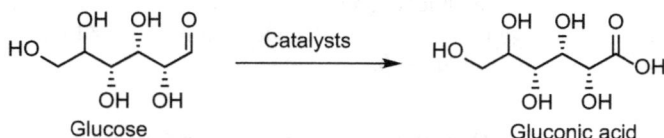

FIGURE 5.1 Conversion of glucose to gluconic acid.

TABLE 5.2
Catalysts Involved during Conversion of Glucose to Gluconic Acid [65–67]

S. No.	Catalyst Used	Oxidant	Solvent	Selectivity (%)
1	Au/Al$_2$O$_3$	O$_2$	Water	97
2	Nanosized Au/SiO$_2$	H$_2$O$_2$	Water	80
3	Pb-Te/SiO$_2$	O$_2$	—	88.4

and toxic for animals [68]. Oxidation of CO is a challenging process; therefore, a highly active oxidation catalyst is required to remove CO efficiently [68b]. For this purpose, Haruta *et al.* revealed that highly dispersed gold prepared on different metal oxide supports are incredibly active in CO oxidation prepared by deposition-precipitation and co-precipitation method, and they observed that the performance of the catalysts was dependent on the mode of the method used for their preparation, and TiO$_2$-supported gold or platinum catalyst exhibits the highest activity.

5.6.8 OXIDATION OF ALKYL-SUBSTITUTED BENZENE

Catalytic oxidation of alkyl-substituted benzene is one of the crucial paths for the synthesis of perfumes, essential chemicals, drugs, and synthetic materials. Different catalysts are employed in the oxidation reactions of alkylbenzene. Table 5.3 summarizes the oxidation of ethylbenzene to acetophenone (Figure 5.2) using different catalysts. The best yield could be realized when Ag/SBA-15 was used as a catalyst and tert-Butyl hydroperoxide (tBuOOH, TBHP) as an oxidant. This reaction is carried out in a solvent-free condition and requires only 5 h for complete conversion.

TABLE 5.3
Catalysts Used for Oxidation of Alkyl Benzene

S. No.	Catalyst Used	Oxidant	Solvent	Reaction time (h)	Selectivity(%)
1	Fe nanocatalysts on surface of SiO$_2$/Al$_2$O$_3$	TBHP	—	24	89
2	Silica-supported cobalt, NHPI	O$_2$	CH$_3$COOH	24	91
3	Au/SBA-15	TBHP	CH$_3$CN	36	93
4	Ag/SBA-15	TBHP	—	5	99

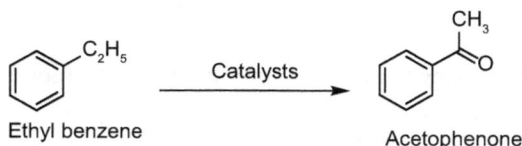

FIGURE 5.2   Conversion of ethyl benzene to acetophenone.

5.6.9   TOTAL OXIDATION OF FORMALDEHYDE

Formaldehyde is harmful and is one of the major pollutants in exhaust gases; hence, catalytic oxidation is the efficient way to control pollution [69]. Among different transition metals like Pt, Pd, Ag, Au, silver is the best metal for oxidation because it is cheap and shows high activity and selectivity [70]. SiO_2-supported silver catalysts favor the oxidation of formaldehyde. Transition metals like Ce and Mn promote the oxidation activity of silver because of high O_2 storage, redox capacity, and oxygen supply to silver (Table 5.4) [71]. An increment in the catalyst activity obtained by the silica modification with cerium or manganese oxide is due to the synergic action between cerium/manganese oxide and silver.

5.6.10   EPOXIDATION OF OLEFINS BY MOLECULAR OXYGEN OVER SUPPORTED METAL HETEROGENEOUS CATALYSTS

Epoxidation of olefins is an effective reaction that yields epoxides as beneficial synthetic intermediates [72]. The molecular oxygen carries out epoxidation of olefins with aldehydes over silica or carbon-supported heterogeneous metal catalysts like Pd/SiO_2, Pd/C, Ru/SiO_2, and Ir/SiO_2. The catalyst is powerful, can be separated easily, provides high selectivity and excellent yield of epoxides. Hence, epoxidation of some common alkenes by molecular oxygen can be effectively carried out by using supported metal heterogeneous catalysts in the presence of aldehydes at room temperature.

Results of epoxidation of isobutyraldehyde or benzaldehyde with different supported metal heterogeneous catalysts at room temperature and 1 atm of O_2 pressure are reported in Table 5.5.

TABLE 5.4
Formaldehyde Conversion and Product Selectivity

S. No.	Sample	HCHO Conversion (%)	Selectivity (methyl formate)
1	CeO_2/SiO_2	19.4	87.5
2	MnO_2/SiO_2	2.9	100
3	$CeO_2\text{-}MnO_x/SiO_2$	8.8	91.7

TABLE 5.5
Epoxidation of Different Olefins in the Presence of Isobutyraldehyde and Benzaldehyde

S. No.	Catalyst	Substrate	Aldehyde	Time (h)	Yield (%)	Selectivity (%)
1	Pd/SiO$_2$	Cyclohexene	Isobutyraldehyde	6.5	98	100
2	Pd/SiO$_2$	1-Hexene	Isobutyraldehyde	21	69	82
3	Pd/SiO$_2$	1-Decene	Benzaldehyde	22	95	100
4	Pd/SiO$_2$	Styrene	Isobutyraldehyde	22	33	41
5	Pd/SiO$_2$	Cyclohexene	Benzaldehyde	12	95	100
6	Pd/C	Cyclohexene	Isobutyraldehyde	5.5	73	99.7
7	Pd/C	1-Decene	Isobutyraldehyde	20	47	100
8	Pd/C	Styrene	Isobutyraldehyde	5	84	84.3
9	Pd/C	1-Decene	Benzaldehyde	22	90	100
10	Ir/SiO$_2$	1-Decene	Isobutyraldehyde	22	40	100
11	Ir/SiO$_2$	1-Decene	Benzaldehyde	22	59	100
12	Pt/SiO$_2$	1-Decene	Isobutyraldehyde	22	23	100
13	Pt/SiO$_2$	1-Decene	Benzaldehyde	22	35.7	100

5.7 TRANSITION METAL COMPLEXES WITH POLYMERIC LIGANDS

Supporting homogeneous transition metal catalysts on organic polymers has been an area of intense research. A considerable number of systems have been studied either using commercially available polymers or by their synthesis. The most popular route to supported catalysts has been to start with commercial polystyrene and introduce functional groups. Polystyrene has been functionalized with a wide variety of ligand groups, and the principle routes include lithiation and chloromethylation. By using polystyrene, the polymer can be modified by changing the amount of cross-linking, affecting the type of catalyst produced. Polystyrene supports can be prepared by chloromethylation of polystyrene cross-linked with divinylbenzene, followed by the treatment with lithium diphenylphosphide (Figure 5.3) [73–76]. However, support has been prepared by polymerizing p-diphenylphosphino styrene. This phosphenated polymer transition metal complex is prepared in an inert solvent.

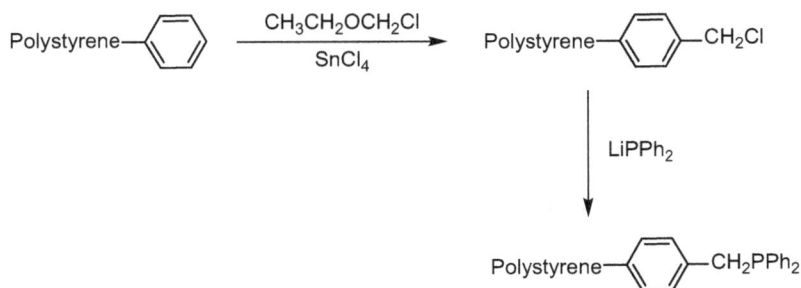

FIGURE 5.3 Preparation of polystyrene support.

Method 1: Preparation of silicon based catalyst on support

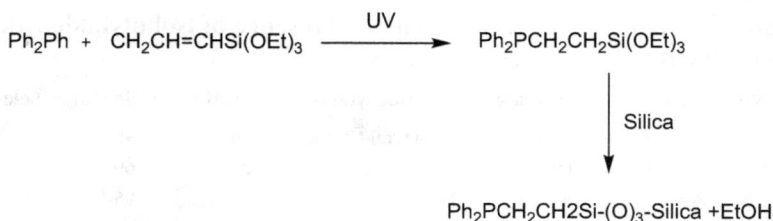

$$Ph_2Ph + CH_2CH=CHSi(OEt)_3 \xrightarrow{UV} Ph_2PCH_2CH_2Si(OEt)_3$$

$$\downarrow \text{Silica}$$

$$Ph_2PCH_2CH2Si\text{-}(O)_3\text{-Silica} + EtOH$$

Method 2: Preparation of metal complex with silica

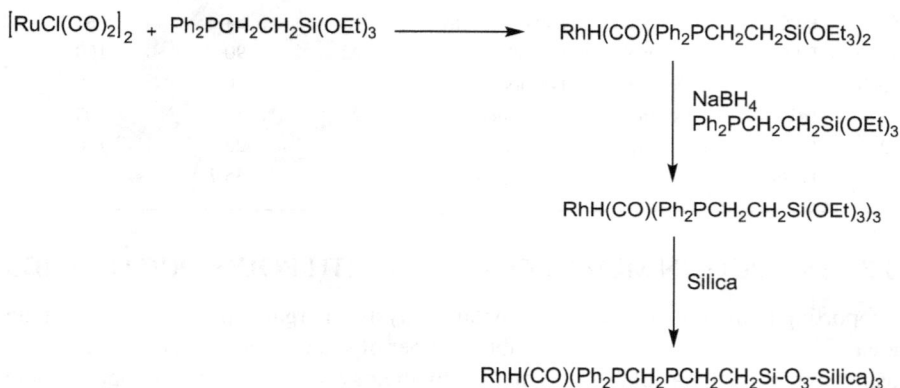

$$\left[RuCl(CO)_2\right]_2 + Ph_2PCH_2CH_2Si(OEt)_3 \longrightarrow RhH(CO)(Ph_2PCH_2CH_2Si(OEt_3)_2$$

$$\downarrow \begin{array}{l} NaBH_4 \\ Ph_2PCH_2CH_2Si(OEt)_3 \end{array}$$

$$RhH(CO)(Ph_2PCH_2CH_2Si(OEt)_3)_3$$

$$\downarrow \text{Silica}$$

$$RhH(CO)(Ph_2PCH_2PCH_2CH_2Si\text{-}O_3\text{-Silica})_3$$

FIGURE 5.4 Method 1: preparation of silicon-based catalyst. Method 2: preparation of metal-based catalyst on silica.

Polymer support formed on silica had been evolved at the British Petroleum Co. Ltd. There are two steps involved in making a transition metal complex with silica ligand [73].

   I. In first method, silica is treated with diphenyl(2-(triethoxysilyl)ethyl)phos-phine to obtain a silicon-based catalyst system which itself is supported on silica. Diphenyl(2-(triethoxysilyl)ethyl)phosphine can be prepared by adding diphenylphosphine to vinyl triethoxysilane under the UV (Figure 5.4) [77].
   II. The second method for preparing silica-supported complex is the initial formation of complex having diphenyl(2-(triethoxysilyl)ethyl)phosphine ligand and condensation with silica (Figure 5.4). For instance, the reaction of ruthenium chloride salt with diphenyl triethoxysilane under ambient reaction conditions leads to the formation of a Ru-complex which is then reduced with $NaBH_4$ and then combined with silica.

There are extensive hydrocarbon reactions that are being catalyzed by silica and polystyrene support mentioned in Table 5.6.

TABLE 5.6
Reactions Catalyzed by Polymer-supported Transition Metals

S. No.	Type of Reaction	Metal Complex	Polymer Support	Substrate	Reference
1	Hydrogenation	$(RhCl(C_2H_4)_2)_2$	Poly-p-diphenylphosphinostyrene	—	[78]
2	Hydrogenation	$RhCl(PPh_3)_3$	Phosphenated 2% cross-linked polystyrene-divinylbenzene	Cyclohexene Hex-1-ene Cyclooctene Cyclododecene	[74]
3	Hydrogenation	$RhCl_3$ $RhCl_3$ then PPh_3 $RhCl_3$ then C_2H_4 $RhCl(PPh_3)_3$	Phosphenated 2% cross-linked polystyrene-divinylbenzene	Hept-1-ene Vinyl acetate Vinyl ethyl ether	[75]
4	Hydrogenation	$NiCl_2$ then $NaBH_4$ $Rh(acac)(CO)_2$ $[RhCl(COD)]_2$ $RhCl_3$	Phosphenated PVC	Propylene Hex-1-ene Oct-1-ene	[79]
5	Hydrogenation	$(EtO)_3SiCH_2$ CH_2PPh_2	Silica	Hex-1-ene	[73]
6	Hydrogenation	$[IrCl(COD)]_2$	Phosphenated silica	Hex-1-ene Isoprene	[80]
7	Hydrogenation	$RhH(CO)(PPh_3)_3$	Phosphenated SiO_2	Pent-1-ene Trans pent-2-ene Cyclohexene Cyclooctene	[81]
8	Hydrogenation	$RhCl_3$	Phosphenated SiO_2	Pent-1-ene	[81]
9	Hydroformylation	$(EtO)_3SiCH_2$ CH_2PPh_2	SiO_2	Hex-1-ene	[73, 82]
10	Hydroformylation	$RhCl_3$ then C_2H_4	Phosphenated 20% cross-linked polystyrene-divinylbenzene	Hept-1-ene	[75]
11	Hydroformylation	$[Rh(CO)_2Cl]_2$	Cross-linked polystyrene-divinylbenzene with- $P(Ph)_2$, $P(Bu)_2$, $-SH$, $-CH_2NMe_2$, $P(OMe)_2$	Hex-1-ene	[83]
12	Hydrosilylation	$RhCl_3$	Phosphenated 20% cross-linked polystyrene-divinylbenzene	$(EtO)_3SiH$ and hex-1-ene or vinyl ethyl ether or acrylonitrile	[75 and 83]
13	Acetoxylation	$PdCl_2$	Phosphenated silica	Ethylene Propylene Isobutene	[84]
14	Polymerization	$Ni(COD)_2$	Phosphenated silica	Butadiene	[80]
15	Polymerization	$NiCl_2$	Phosphenated polystyrene	Ethyl propiolate	[84]
16	Oligomerization	$NiCl_2$	Phosphenated polystyrene	Phenyl acetylene	[84]

REFERENCES

1. Gates, B.C. (1992). *Catalytic Chemistry*. Wiley, New York.
2. Satterfield, C. N. (1991). *Heterogeneous Catalysis in Industrial Practice*. McGraw-Hill, New York.
3. Farrauto, R. J., & Bartholomew, C. H. (1997). *Fundamentals of Industrial Catalytic Processes*. Blackie Academic and Professional, London.
4. Ertl, G., & Knözinger, H. (1997). *Handbook of Heterogeneous Catalysis*, Vols. 1–5. In J. Weitkamp (Ed.). Wiley-VCH, Weinheim.
5. Hagen, J. (1999). *Industrial Catalysis, A Practical Approach*. Wiley-VCH, Weinheim.
6. Yoon, M., Srirambalaji, R., & Kim, K. (2011). Homochiral metal–organic frameworks for asymmetric heterogeneous catalysis. *Chemical Reviews*, *112*(2), 1196–1231.
7. Corma, A., García, H., & Llabrés i Xamena, F. X. (2010). Engineering metal organic frameworks for heterogeneous catalysis. *Chemical Reviews*, *110*(8), 4606–4655.
8. Kajbafvala, A., Zanganeh, S., Kajbafvala, E., Zargar, H. R., Bayati, M. R., & Sadrnezhaad, S. K. (2010). Microwave-assisted synthesis of narcis-like zinc oxide nanostructures. *Journal of Alloys and Compounds*, *497*(1–2), 325–329.
9. Kumar, A., Kumar, V. P., Kumar, B. P., Vishwanathan, V., & Chary, K. V. R. (2014). Vapor phase oxidation of benzyl alcohol over gold nanoparticles supported on mesoporous TiO_2. *Catalysis Letters*, *144*(8), 1450–1459.
10 Burri, D. R., Shaikh, I. R., Choi, K. M., & Park, S. E. (2007). Facile heterogenization of homogeneous ferrocene catalyst on SBA-15 and its hydroxylation activity. *Catalysis Communications*, *8*(4), 731–735.
11. Sreevardhan Reddy, S., David Raju, B., Siva Kumar, V., Padmasri, A. H., Narayanan, S., & Rama Rao, K. S. (2007). Sulfonic acid functionalized mesoporous SBA-15 for selective synthesis of 4-phenyl-1,3-dioxane. *Catalysis Communications*, *8*(3), 261–266.
12. Kim, D. J., Dunn, B. C., Cole, P., Turpin, G., Ernst, R. D., Pugmire, R. J., Kang, M., Kim, J. M., & Eyring, E. M. (2005). Enhancement in the reducibility of cobalt oxides on a mesoporous silica supported cobalt catalyst. *Chemical Communications*, *(11)*, 1462–1464.
13. Burri, D. R., Jun, K. W., Kim, Y. H., Kim, J. M., Park, S. E., & Yoo, J. S. (2002). Cobalt catalyst heterogenized on SBA-15 for p-xylene oxidation. *Chemistry Letters*, *31*(2), 212–213.
14. Anand, N., Reddy, K. H. P., Prasad, G. V. S., Rama Rao, K. S., & Burri, D. R. (2012). Selective benzylic oxidation of alkyl substituted aromatics to ketones over Ag/SBA-15 catalysts. *Catalysis Communications*, *23*, 5–9.
15. Nam, J. H., Jang, Y. Y., Kwon, Y. U., & Nam, J. D. (2004). Direct methanol fuel cell Pt–carbon catalysts by using SBA-15 nanoporous templates. *Electrochemistry Communications*, *6*(7), 737–741.
16. Arsalanfar, M., Mirzaei, A. A., Bozorgzadeh, H. R., Samimi, A., & Ghobadi, R. (2014). Effect of support and promoter on the catalytic performance and structural properties of the Fe–Co–Mn catalysts for Fischer–Tropsch synthesis. *Journal of Industrial and Engineering Chemistry*, *20*(4), 1313–1323.
17. Kajbafvala, A., Shayegh, M. R., & Mazloumi, M. (2009). Nanostructure sword-like ZnO wires: rapid synthesis and characterization through a microwave-assisted route. *Journal of Alloys and Compounds*, *469*, 293–297.
18. Ali, M. E., Rahman, M. M., Sarkar, S. M., & Hamid, S. B. A. (2014). Heterogeneous metal catalysts for oxidation reactions. *Journal of Nanomaterials*, *2014*, 1–23.
19. Courty, P., & Marcilly, C. (1983). *Preparation of Catalysts III*. In G. Poncelet, P. Grange, P. Jacobs (Eds.), (p. 485). Elsevier, Amsterdam.

20. Stiles, B., & Koch, T. A. (1995). *Catalyst Manufacture*, 2nd. ed. In M. Dekker (Ed.), (p. 291). Marcel Dekker Inc., New York.
21. Geus, J. W., & Dillen, J. V. (1997). Handbook of Heterogeneous Catalysis, Vol. 1. *Preparation of Supported Catalysts by Deposition–Precipitation*. In G. Ertl, H. Knözinger, & J. Weitkamp (Eds.), (p. 240). Wiley-VCH, Weinheim.
22. Ledoux, M. J., & Pham-Huu, C. (2001). *CATTECH*, *5(4)*, 226–246.
23. Sieber, H., Hoffmann, C., Kaindl, A., & Greil, P. (2000). Biomorphic cellular ceramics. *Advanced Engineering Materials*, *2(3)*, 105–109.
24. Knözinger, H., & Taglauer, E. (1997). *Handbook of Heterogeneous Catalysis*, Vol. 1. In G. Ertl, H. Knözinger, & J. Weitkamp (Eds.), (p. 216). Wiley-VCH, Weinheim.
25. Wachs, I. E. (1997). *Catalysis*, Vol. 13, The Royal Society of Chemistry, Cambridge.
26. Delmon, B. (1997). *Handbook of Heterogeneous Catalysis*, Vol. 1. In G. Ertl, H. Knözinger, & J. Weitkamp (Eds.), (p. 264). Wiley-VCH, Weinheim.
27. Centi, G. (1996). Nature of active layer in vanadium oxide supported on titanium oxide and control of its reactivity in the selective oxidation and ammoxidation of alkylaromatics. *Applied Catalysis A: General*, *147(2)*, 267–298.
28. Janssen, F. J. (1997). *Handbook of Heterogeneous Catalysis*, Vol. 4. In G. Ertl, H. Knözinger, & J. Weitkamp (Eds.), (p. 1633). Wiley-VCH, Weinheim.
29. Armor, J. N. (1994). Materials needs for catalysts to improve our environment. *Chemistry of Materials*, *6(6)*, 730–738.
30. Mol, J. C. (1997). *Handbook of Heterogeneous Catalysis*, Vol. 5. In G. Ertl, H. Knözinger, & J. Weitkamp (Eds.), (p. 2387). Wiley-VCH, Weinheim.
31. Buonomo, F., Sanfilippo, D., & Trifiro, F. (1997). *Handbook of Heterogeneous Catalysis*, Vol. 5. In G. Ertl, H. Knözinger, & J. Weitkamp (Eds.), (p. 2140). Wiley-VCH, Weinheim.
32. Overbury, S. H., Bertrand, P. A., & Somorjai, G. A. (1975). Surface composition of binary systems. Prediction of surface phase diagrams of solid solutions. *Chemical Reviews*, *75(5)*, 547–560.
33. Che, M., Clause, O., & Marcilly Ch. (1997). *Handbook of Heterogeneous Catalysis*, Vol. 1. In G. Ertl, H. Knözinger, & J. Weitkamp (Eds.), (p. 191), Wiley-VCH, Weinheim.
34. Boehm, H. P., & Knözinger, H. (1983). *Catalysis: Science and Technology*, Vol. 4. In J. R. Anderson, & M. Boudart (Eds.), (p. 39). Springer, Berlin.
35. Ono, Y., & Baba, T. (2000). *Catalysis*, Vol. 15, (p. 1). The Royal Society of Chemistry, Cambridge.
36. Tanabe, K., & Hattori, H. (1997). *Handbook of Heterogeneous Catalysis*, Vol. 1. In G. Ertl, H. Knözinger, & J. Weitkamp (Eds.), (p. 404). Wiley-VCH, Weinheim.
37. Olah, G. A., & Sommer, J. (1998). *Topics in Catalysis*, Vol. 6. Superacid catalysis. Baltzer Science Publishers, Amsterdam, Netherlands.
38. Duchet, J. (1983). Carbon-supported sulfide catalysts. *Journal of Catalysis*, *80(2)*, 386–402.
39. De Beer, V. H. J., Van der Aalst, M. J. M., Machiels, C. J., & Schuit, G. C. A. (1976). The CoO MoO₃ γ-AL₂O₃ catalyst VII. Influence of the support. *Journal of Catalysis*, *43(1–3)*, 78–89.
40. De Beer, V. H. J., & Schuit, G. C. A. (1976). *Preparation of Catalysts*. In B. Delmon, P. A. Jacobs, & G. Poncelet (Eds.), (p. 343). Elsevier, Amsterdam.
41. Foger, K. (1984). *Catalysis: Science and Technology*, Vol. 6. In J. R. Anderson, & M. Boudart (Eds.), (p. 228). Springer, Berlin.
42. Baiker, A. (1998). Chiral catalysis on solids. *Current Opinion in Solid State and Materials Science*, *3(1)*, 86–93.

43. Vos, D. E. D., Vankelecom, I. F. J., & Jacobs, P. A. (2000). *Chiral Catalyst Immobilization and Recycling*, Wiley-VCH, Weinheim.
44. Bosman, A. W., Janssen, H. M., & Meijer, E. W. (1999). About dendrimers: structure, physical properties, and applications. *Chemical Reviews*, *99(7)*, 1665–1688.
45. (a) De Vos, D. E., Sels, B. F., & Jacobs, P. A. (2001). Immobilization of homogeneous oxidation catalysts. *Advances in Catalysis*, *46*, 1–87; (b) Clapham, B., Reger, T., & Janda, K. D. (2001). Polymer-supported catalysis in synthetic organic chemistry. *Tetrahedron*, *57*, 4637–4662; (c) Kirschning, A., Monenschein, H., & Wittenberg, R. (2001). Functionalized polymers-emerging versatile tools for solution-phase chemistry and automated parallel synthesis. *Angewandte Chemie International Edition*, *40(4)*, 650–679.
46. Ertl, G., Knözinger, H. & Weitkamp, J. (2008). *Handbook of Heterogeneous Catalysis*. Wiley-VCH, GmbH & Co., KGaA.
47. (a) Sirisha, V. L., Jain, A., & Jain, A. (2016). Enzyme Immobilization. *Advances in Food and Nutrition Research*, *79*, 179–211; (b) Vennestrøm, P. N. R., Christensen, C. H., Pedersen, S., Grunwaldt, J. D. & Woodley, J. M. (2010). Next-generation catalysis for renewables: combining enzymatic with inorganic heterogeneous catalysis for bulk chemical production. *ChemCatChem*, *2*, 249–258.
48. Engström, K., Johnston, E. V., Verho, O., Gustafson, K. P. J., Shakeri, M., Tai, C. W., & Bäckvall, J. E. (2013). co-immobilization of an enzyme and a metal into the compartments of mesoporous silica for cooperative tandem catalysis: an artificial metalloenzyme. *Angewandte Chemie International Edition*, *52(52)*, 14006–14010.
49 (a) Ekloff, G. S., & Ernst, S. (1997). *Handbook of Heterogeneous Catalysis*, Vol. 1. In G. Ertl, H. Knözinger, & J. Weitkamp (Eds.), (p. 374). Wiley-VCH, Weinheim; (b) Vos, D. E. D., Gerrits, P. P. K., Parton, R. F., Weckhuysen, B. M., Jacobs, P. A., & Schoonheydt, R. A. (1995). Coordination chemistry in zeolites. *Journal of Inclusion Phenomena and Molecular Recognition in Chemistry*, *21*, 185.
50. (a) McDaniel, M. P. (1997). *Handbook of Heterogeneous Catalysis*, Vol. 5. In G. Ertl, H. Knözinger, & J. Weitkamp (Eds.), (p. 2400). Wiley-VCH, Weinheim; (b) Kaminsky, W. (2001). Olefin polymerization catalyzed by metallocenes. *Advances in Catalysis*, *46*, 89–159.
51. Pangarkar, K., Schildhauer, T. J., van Ommen, J. R., Nijenhuis, J., Kapteijn, F., & Moulijn, J. A. (2008). structured packings for multiphase catalytic reactors. *Industrial & Engineering Chemistry Research*, *47(10)*, 3720–3751.
52. Twigg, M. V., & Richardson, J. T. (2007). Fundamentals and applications of structured ceramic foam catalysts. *Industrial & Engineering Chemistry Research*, *46(12)*, 4166–4177.
53. Matatov-Meytal, Y., & Sheintuch, M. (2002). Catalytic fibers and cloths. *Applied Catalysis A: General*, *231(1–2)*, 1–16.
54. (a) Liang, G., He, L., & Cheng, H. (2014). The hydrogenation/dehydrogenation activity of supported Ni catalysts and their effect on hexitols selectivity in hydrolytic hydrogenation of cellulose. *Journal of Catalysis*, *309*, 468–476; (b) Luo, Q., Beller, M., & Jiao, H. (2013). Formic acid dehydrogenation on surfaces-A Review of computational aspect. *Journal of Theoretical and Computational Chemistry*, *12(7)*, 1–28.
55 (a) Palcheva, R., Dimitrov, L., Tyuliev, G., Spojakina, A., & Jiratova, K. (2013). TiO_2 nanotubes supported NiW hydrodesulphurization catalysts: characterization and activity. *Applied Surface Science*, *265*, 309–316; (b) Bagheri, S., Shameli, K., & Hamid, S. B. A. (2013). Synthesis and characterization of anatase titanium dioxide nanoparticles using egg white solution via Sol-Gel method. *Journal of Chemistry*, *2013*, 1–5; (c) Kominami, H., Kato, J. I., Takada, Y., Doushi, Y., Ohtani, B., Nishimoto, S., Inoue,

M., Inui, T., & Kera, Y. (1997). Novel synthesis of microcrystalline titanium (IV) oxide having high thermal stability and ultra-high photocatalytic activity: thermal decomposition of titanium (IV) alkoxide in organic solvents. *Catalysis Letters, 46 (1–2)*, 235–240.

56 (a) Bamwenda, G. R., Tsubota, S., Nakamura, T., & Haruta, M. (1997). The influence of the preparation methods on the catalytic activity of platinum and gold supported on TiO$_2$ for CO oxidation. *Catalysis Letters, 44, (1–2)*, 83–87; (b) Tauster, S. J., Fung, S. C., Baker, R. T. K., & Horsley, J. A. (1981). Strong interactions in supported-metal catalysts. *Science, 211*, 1121–1125.

57. Nolan, M. (2013). Modifying ceria (111) with a TiO$_2$ nanocluster for enhanced reactivity. *Journal of Chemical Physics, 139 (18)*, 1–8.

58 (a) Carrettin, S., McMorn, P., Johnston, P., Griffin, K., & Hutchings, G. J. (2002). Selective oxidation of glycerol to glyceric acid using a gold catalyst in aqueous sodium hydroxide. *Chemical Communications*, (7), 696–697; (b) Porta, F., Prati, L., Rossi, M., Coluccia, S., & Martra, G. (2000). Metal sols as a useful tool for heterogeneous gold catalyst preparation: Reinvestigation of a liquid phase oxidation. *Catalysis Today, 61 (1)*, 165–172; (c) Fang, W., Chen, J., Zhang, Q., Deng, W., & Wang, W. (2011). Hydrotalcite supported gold catalyst for the oxidant-free dehydrogenation of benzyl alcohol: studies on support and gold size effects. *Chemistry—A European Journal, 17 (4)*, 1247–1256.

59. Kim, T. S., Stiehl, J. D., Reeves, C. T., Meyer, R. J., & Mullins, C. B. (2003). Cryogenic CO oxidation on TiO$_2$-supported gold nanoclusters precovered with atomic oxygen. *Journal of the American Chemical Society, 125(8)*, 2018–2019.

60. Li, G., Enache, D. I., Edwards, J., Carley, A. F., Knight, D. W., & Hutchings, G. J. (2006). Solvent-free oxidation of benzyl alcohol with oxygen using zeolite-supported Au and Au–Pd catalysts. *Catalysis Letters, 110(1–2)*, 7–13.

61 (a) Shinde, V. M., & Madras, G. (2013). CO methanation toward the production of synthetic natural gas over highly active Ni/TiO$_2$ catalyst. *AIChE Journal, 60(3)*, 1027–1035; (b) Ullah, K., Ye, S., Sarkar, S., Zhu, L., Meng, Z. D., & Oh, W. C. (2014). Photocatalytic degradation of methylene blue by NiS$_2$-graphene supported TiO$_2$ catalyst composites. *Asian Journal of Chemistry, 26(1)*, 145–150; (c) Song, H., Dai, M., Wan, X., Xu, X., Zhang, C., & Wang, H. (2014). Synthesis of a Ni$_2$P catalyst supported on anatase-TiO$_2$ whiskers with high hydrodesulfurization activity, based on triphenylphosphine. *Catalysis Communications, 43*, 151–154.

62. Kong, W., Zhang, X. H., Zhang, Q., Wang, T. J., Ma, L. L., & Chen, G. Y. (2013). Hydrodeoxygenation of guaiacol over nickelbased catalyst supported on mixed oxides. *Chemical Journal of Chinese Universities, 34 (12)*, 2806–2813.

63. Morales, F., Smit, E. D., de Groot, F. M. F., Visser, T., & Weckhuysen, B. M. (2007). Effects of manganese oxide promoter on the CO and H$_2$ adsorption properties of titania-supported cobalt Fischer-Tropsch catalysts. *Journal of Catalysis, 246(1)*, 91–99.

64. (a) Hemalatha, K., Madhumitha, G., Kajbafvala, A., Anupama, N., Sompalle, R., & Roopan, S. M. (2013). Function of nanocatalyst in chemistry of organic compounds revolution: an overview. *Journal of Nanomaterials, 2013*, 1–23; (b) Kropp, P. J., Breton, G. W., Fields, J. D., Tung, J. C., & Loomis, B. R. (2000). Surface-mediated reactions. 8. Oxidation of sulfides and sulfoxides withtert-butyl hydroperoxide and oxone. *Journal of the American Chemical Society, 122(18)*, 4280–4285; (c) Biradar, A. V., & Asefa, T. (2012). Nanosized gold-catalyzed selective oxidation of alkyl-substituted benzenes and n-alkanes. *Applied Catalysis A: General, 435–436*, 19–26.

65. Delidovich, I. V., Moroz, B. L., Taran, O. P., Gromov, N. V., Pyrjaev, P. A., Prosvirin, I. P., Bhuktiyarov, V. I., & Parmon, V. N. (2013). Aerobic selective oxidation of glucose to gluconate catalyzed by Au/Al$_2$O$_3$ and Au/C: Impact of the mass-transfer processes on the overall kinetics. *Chemical Engineering Journal, 223*, 921–931.

66. Bujak, P., Bartczak, P., & Polanski, J. (2012). Highly efficient room-temperature oxidation of cyclohexene and D-glucose over nanogold Au/SiO$_2$ in water. *Journal of Catalysis, 295,* 15–21.

67. Witońska, I., Frajtak, M., & Karski, S. (2011). Selective oxidation of glucose to gluconic acid over Pd–Te supported catalysts. *Applied Catalysis A: General, 401(1–2),* 73–82.

68. (a) Kirichenko, O. A., Redina, E. A., Davshan, N. A., Mishin, I. V., Kapustin, G. I., Brueva, T. R., Kustov, L. M., Li, W., & Kim, C. H. (2013). Preparation of alumina-supported gold-ruthenium bimetallic catalysts by redox reactions and their activity in preferential CO oxidation. *Applied Catalysis B: Environmental, 134–135,* 123–129; (b) Choudhary, T. V., Sivadinarayana, C., Chusuei, C. C., Datye, A. K., Fackler Jr., J. P., & Goodman, D. W. (2002). CO oxidation on supported nano-Au catalysts synthesized from a [Au$_6$(PPh$_3$)$_6$](BF4)$_2$ complex. *Journal of Catalysis, 207* (2), 247–255.

69 (a) Sekine, Y. (2002). Oxidative decomposition of formaldehyde by metal oxides at room temperature. *Atmospheric Environment, 36(35),* 5543–5547; (b) Luo, M., Yuan, X., & Zheng, X. (1998). Catalyst characterization and activity of Ag–Mn, Ag–Co and Ag–Ce composite oxides for oxidation of volatile organic compounds. *Applied Catalysis A: General, 175(1–2),* 121–129.

70 (a) Tang, X., Chen, J., Li, Y., Li, Y., Xu, Y., & Shen, W. (2006). Complete oxidation of formaldehyde over Ag/MnO$_x$–CeO$_2$ catalysts. *Chemical Engineering Journal, 118(1–2),* 119–125; (b) Imamura, S., Yamada, H., & Utani, K. (2000). Combustion activity of Ag/CeO$_2$ composite catalyst. *Applied Catalysis A: General, 192(2),* 221–226; (c) Hamoudi, S., Sayari, A., Belkacemi, K., Bonneviot, L., & Larachi, F. (2000). Catalytic wet oxidation of phenol over PtxAg$_{1-x}$MnO$_2$/CeO$_2$ catalysts. *Catalysis Today, 62(4),* 379–388; (d) Scirè, S., Crisafulli, C., Giuffrida, S., Mazza, C., Riccobene, P. M., Pistone, A., Ventimiglia, G., Bongiorno, C., & Spinella, C. (2009). Supported silver catalysts prepared by deposition in aqueous solution of Ag nanoparticles obtained through a photochemical approach. *Applied Catalysis A: General, 367(1–2),* 138–145; (e) Scirè, S., Riccobene, P. M., & Crisafulli, C. (2010). Ceria supported group IB metal catalysts for the combustion of volatile organic compounds and the preferential oxidation of CO. *Applied Catalysis B: Environmental, 101(1–2),* 109–117; (f) Ye, Q., Zhao, J., Huo, F., Wang, J., Cheng, S., Kang, T., & Dai, H. (2011). Nanosized Ag/α-MnO$_2$ catalysts highly active for the low-temperature oxidation of carbon monoxide and benzene. *Catalysis Today, 175(1),* 603–609.

71. Kharlamova, T., Mamontov, G., Salaev, M., Zaikovskii, V., Popova, G., Sobolev, V., Knyazev, A., & Vodyankina, O. (2013). Silica-supported silver catalysts modified by cerium/manganese oxides for total oxidation of formaldehyde. *Applied Catalysis A: General, 467,* 519–529.

72 (a) Weisermel, K., & Atpe, H. J. (1993). *Industrial Organic Chemistry,* VCH, Weinheim; (b) House, H. (1972). *Modern Synthetic Reactions.* In W. A. Benjamin (Ed.), (pp. 292–321). Elsevier, New York.

73. Allurn, K. G., Hancock, R. D., McKenzie, S., & Pitkethly, R. C. (1972). *Proc. 5th Znternat. Cong. Catalysis, Palm Beach.*

74. Grubbs, R. H., & Kroll, L. C. (1971). Catalytic reduction of olefins with a polymer-supported rhodium(I) catalyst. *Journal of the American Chemical Society, 93(12),* 3062–3063.

75. Capka, M., Svoboda, P., Kraus, M., & Hetflejs, J. (1971). *Tetrahedron Letters, 50(12),* 4787–4790.

76. B.P. Co. Ltd. (1972). *British Patent* 1277737.

77. Niebergall, H. (1962). *Makromol. Chenz.,* 52, 218

78. Manassen, J. (1974). Modification of the redox-properties of tetraphenylporphyrin-complexes by bases in methylene-chloride solution. The equilibrium of different oxidation-states with added base as measured by cyclic voltammetry. *Israel Journal of Chemistry*, *12*(6), 1059–1067.
79. B.P. Co. Ltd. (1972). *British Patent* 1295475.
80. B.P. Co. Ltd. (1973). *U.S. Patent* 3726809.
81. Michalska, Z. M., & Webster, D. E. (1974). Supported homogeneous catalysts, *Platinum Metals Review*, *18*(2), 65–73.
82. B.P. Co. Ltd. 1970. *Dutch Patent* 7006740.
83. Svoboda, P., Capka, M., Chvalovsky, V., Bazant, J., Hetflejs, H., & Pracejus, H. (1972). *Angewandte Chemie*, *12*, 153.
84. B.P. Co. Ltd. (1972). *British Patent* 1295674.

6 Mesoporous Materials in Heterogeneous Catalysis

Meenal Batra
Banasthali Vidyapith, Newai (Rajasthan), India

Ashutosh Sharan Singh
Maharishi Markandeswar Deemed to be University,
Haryana, India

CONTENTS

6.1 INTRODUCTION

Molecular pores or voids provide a confined environment than the bulk phase. These pores may be constrained or flexible depending upon the wall of molecular architecture and target molecule of consideration. Zeolites [1] are natural porous architectures and scientists have attempted to mimic such pores in an artificial system which may be synthesized either through coordinate bond between metal cations (acts as node) and judicial choice of an organic framework (known as linker) or through condensation of purely organic framework. The former is explored as metal–organic frameworks (MOFs) [2] and the latter is known as covalent organic frameworks (COFs) [3]. Depending upon the range of pore size, materials are classified as microporous, mesoporous and macroporous, respectively. Materials of pore size in the range of 2–50 nm are known as mesoporous materials. The word "Mesoporous" is of Greek origin, which signifies having a pore size of diameter in between micro (less than 2 nm) and macro (more than 50 nm) materials.

DOI: 10.1201/9781003126270-6

6.2 TYPES AND CLASSIFICATION OF MESOPOROUS MATERIALS

There are other types of porous materials, namely porous organic polymers (POPs) [4], porous molecular solids (PMS, Table 6.1). Every category has its own significance and suitable applications. Such porous materials have a wide range of applications like gas storage [5–12], separations [13–21], sensing [22–26], magnetism [27–33], catalysis [34–41], etc. In this chapter, we will focus on various catalytic processes under heterogeneous conditions and their underlying basic principles.

6.3 CHARACTERISTIC FEATURES OF MESOPOROUS MATERIALS FOR CATALYSIS

Molecular pores or cavities of mesoporous materials play a significant role in the chemical transformation [42]. The pore size of mesoporous materials plays a dual role: (i) acts as a barrier for mass transfer, and (ii) provides molecular exchange resistivity. The prerequisite conditions for mesoporous materials to qualify as an ideal system for catalysis with a high turnover number (TON) are *rigidity* and *insolubility*. The rigidity of a molecular pore stabilizes the transition-state and determines the selectivity, observed in the final product. Isolated molecular pore even shows some unusual chemical transformation that may not occur in an open system [43]. For catalytic purposes, porous materials should qualify the following characteristic features:

   (i)   it should possess suitable pore size, should be easily accessible and if required can be programmed accordingly,
  (ii)   should be chemically and thermally stable during the course of reaction,
 (iii)   should be reusable, and
 (iv)   it should be overall cost-effective.

Among mesoporous materials, MOFs meet all these criteria. An additional advantage of using MOFs as mesoporous materials for catalytic purposes is their ease of tuning just by changing the ratio of metal composition (nodes) and/or their organic

TABLE 6.1
Classification of Porous Materials and Their Characteristic features

	Zeolites	MOFs	COFs	POPs	PMS
Porosity	Microporous or mesoporous	Mesoporous	Mesoporous	Mostly microporous	Mesoporous
Crystallinity	High	High	Moderate	Amorphous	High/ amorphous
Thermal stability	Excellent	Good	Poor	Okay	Poor
Modularity/ diversity	High	Excellent	High	High	Possible
Processing	Insoluble	Insoluble	Insoluble	Moderate	Soluble
Designability	Excellent	Excellent	Good	Good	Possible

counterparts (linkers). Various types of organic linkers have been explored in literature, such as phenolate, pyridyl, amine, imidazolates, sulphonates, phosphonates, polycarboxylates, etc. It has been found that the organic linker-bearing carboxylate group with transition metal cations form stable and robust pores like zeolites [44].

For chemical transformation, the surrounding medium around participating molecule(s) plays a crucial role; for example, chemical transformation systems that are uniform or homogeneous in nature, perform well, take less time to complete the reaction and hence, utilized in industrial processes [45]. However, the isolation of catalysts and their recycling is a highly tedious and challenging process [46]. Mesoporous materials, especially MOFs work as heterogeneous systems and facilitate several advantages like [46–48]:

 (i)   ease of separation (after completion of a reaction),
 (ii) easy recycling,
 (iii) minimisation of metal contamination in the product(s),
 (iv) improved handling and process control,
 (v)   single-site reactivity,
 (vi) pore-defined substrate size and shape selectivity, etc.

The efficiency, selectivity and loyalty of heterogeneous catalysis through mesoporous materials mainly depend upon the composition and nature of pore(s) of materials, i.e., linker(s) and node(s) utilized in the formation of mesoporous materials. Various linkers containing different functional groups or sometimes mixed functional groups have been explored in the literature. Preference has been given to linkers containing carboxylic acid group, because of the thermal stability of the metal–ligand bonding of the resultant architecture [44].

6.4 BUILDING BLOCKS OF MOFS

The fate of physicochemical properties, like thermal stability, size, and shape of the pore formed in a 3-D MOF network, depends upon the building blocks of the MOF. Hence, to construct a MOF, a judicial choice of building block is very crucial. There are mainly two parts: the first is organic moiety, also called a linker and the second is an inorganic component, called a node. Depending upon our requirement, i.e., shape and size of the pore, various linkers of different shapes and sizes are explored in the literature (as shown below). The nature and position of the functional group at the organic linker also play a decisive role. In general, a single type of linker is used to construct the corresponding MOF. However, it has been observed that mixed linkers give extra strength to the framework that we will see in various examples, discussed below. Depending upon the number of binding sites (to node), the linkers can be classified as monodentate, didentate, tridentate, and so on.

Another factor that controls the shape and size of pore(s) in mesoporous materials is the node. In fact, the node of MOFs plays multidisciplinary roles: (i) it determines rigidity/flexibility of the resultant frameworks, (ii) it determines shape and size of pore(s) formed in the resultant frameworks, (iii) it provides a catalytic site for chemical transformation. In general, if the coordination site of a metal ion is occupied by solvent molecule(s) then it can be removed either by heating, applying high pressure or through diffusion. The resultant vacant site is occupied later by the active site of the reactant molecule in the catalytic process. To incorporate the active site on the metal ion, metal of high coordination number, such as lanthanides or actinides have been successfully explored (that will be discussed later in catalytic methodologies). The concepts of secondary building units (SBUs) [49–51] and isoreticular synthesis [49] have been used to obtain structural diversity in MOFs for several applications including heterogeneous catalysis. In this chapter, we will focus on the basic concepts and applications of heterogeneous catalysis rather than sequential improvement or development of MOFs formation, which is already well explored [5, 49, 50, 52–56] in several reviews.

6.5 BASIC UNDERLYING PRINCIPLES OF HETEROGENEOUS CATALYSIS WITH MOFS

In general, MOFs synthesized either by hydrothermal or solvothermal methodology provide a 3-D architecture (in contrary to room temperature synthesis which form 2-D sheet) with a periodic arrangement of interspatial pores or cavities. Also, resultant frameworks are insoluble in common organic solvents and thus facilitate heterogeneous catalysis. In a typical MOF, the following catalytic sites are available for chemical transformations:

(i)   *Unsaturated metal centre or coordinatively unsaturated site (CUS)*,
(ii)   *Surface modification or functionalisation of linker unit*, and
(iii) *central pore/cavity of MOFs*.

The last two methodologies have been developed recently for novel and robust MOF formation for multidisciplinary applications including catalysis. Catalytic processes on each site with some selected examples will be discussed.

6.5.1 UNSATURATED METAL CENTRE

During the course of framework synthesis, if the coordination number of the metal centre is fulfilled by coordinated solvent molecule(s) such as water, DMF, methanol, ethanol, etc., then the solvent molecule(s) may be removed either by heating or applying high vacuum. Sometimes, the framework collapses after the removal of the coordinated solvent molecule(s) by heating MOFs under high pressure. To avoid the framework collapse, two methods have been explored, called as supercritical CO_2 treatment [57] and freeze-drying method [58]. The coordinated solvent molecule(s) removed by either method provides a catalytic active site (called as coordinatively unsaturated site, CUS, Figure 4) at metal ion. To create such an active site, transition metal ion of high coordination number or lanthanide metal ions are used. These metal ions have high chances to fulfil the coordination site by surrounding solvent molecule(s) during the process of their synthesis. Chromium terephthalate-based MOF, MIL-101 reported by G. Férey [59], was stable at 275 °C even after removing coordinated water molecules. It was possible to replace water molecules attached to the metal centre of the resulting MOF with suitable chiral auxiliary for asymmetric catalysis [60] (*vide infra*).

The first example of heterogeneous catalysis at the metal centre of a crystalline porous material has been explored by Fujita et al. [61]. They have shown size-selective cyanosilylation of aldehydes. The metal ion at the node of MOF was facilitating Lewis acidic sites for catalytic process. The application of a free basic site at the node of MOF has been explored by Kim et al. for transesterification process through POST-1 (Figure 6.1) [62]. The carboxylate group of ligands (D-PTT, shown above) binds with Zn^{2+} ion and N-terminal pyridyl group remains free and this free basic site catalyses transesterification reaction.

Asymmetric catalysis at the node of MOF was first explored by Lin and co-workers. As discussed above, to create a CUS at metal ion, it is worthful to use metal ion of high coordination number. Lin and co-workers successfully synthesized MOF by using one isomer of 2,2'-diethoxy-1,1'-binaphthalene-6,6'-bisphosphonic acid with a series of lanthanide ions to get crystals of homochiral lanthanide bisphosphonate. Enantioselective cyanosilylation of benzaldehyde was performed [63] using samarium derivative of MOF (Figure 6.2). Although selectivity was less, enhancement of selectivity was explored through a novel approach.

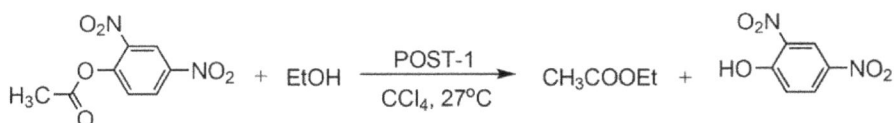

FIGURE 6.1 Free basic site attached to chiral MOF (POST-1) for *trans*-esterification process.

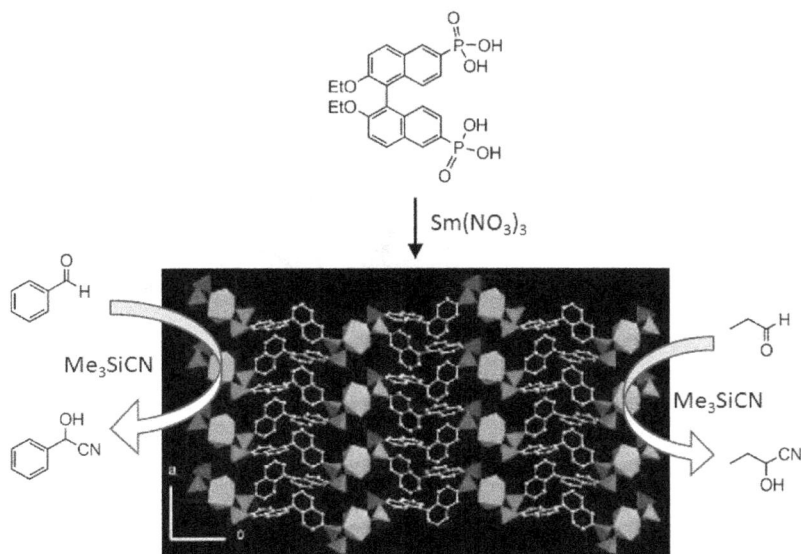

FIGURE 6.2 First asymmetric catalysis promoted at the node of MOF, reported by Lin and co-workers.

(Reproduced after permission from reference 63.)

After removing the coordinated solvent molecule, the metal ion at the node acts as Lewis acidic site, so the active site of the reactant molecule coordinates with the Lewis acidic site. Chiral linker provides an asymmetric environment or space for the catalytic process. Catalytic oxidation of substrate is also possible at the node of MOF [64]. Chemoselective oxidation of sulphide to sulphoxide using urea hydroperoxide (UHP) or H_2O_2 was explored [65] by Fedin, Kim and co-workers. MOF, [Zn_2(bdc) (l-lac)(dmf), dmf = N,N-dimethylformamide] was synthesized by using mixed linkers, 1,4-dicarboxylic acid and L-lactic acid with $Zn(NO_3)_2 \cdot 6H_2O$. The conversion and selectivity of catalytic product were dependent upon size of the substrate molecule (Figure 6.3). Although conversion of small substrate (a and b of Table 6.2) was moderate, selectivity was good. However, conversion of the larger substrate was poor. This result shows the significance of the pore size inside MOF.

The perquisite condition for catalysis at the node is to sustain framework integrity after the removal of the coordinated solvent molecule. Among transition metal ions, chromium-based MOF (MIL-101) [59] has shown good thermal stability. Férey and co-workers have explored MIL-101, for catalytic synthesis of sulphoxide [66] simply using hydrogen peroxide (Figure 6.4). Highly effective conversion and excellent selectivity were observed, even for a larger substrate with MOF, MIL-101, of exceptionally large pore size and high surface area. The catalytic process was expected to proceed through a free radical mechanism.

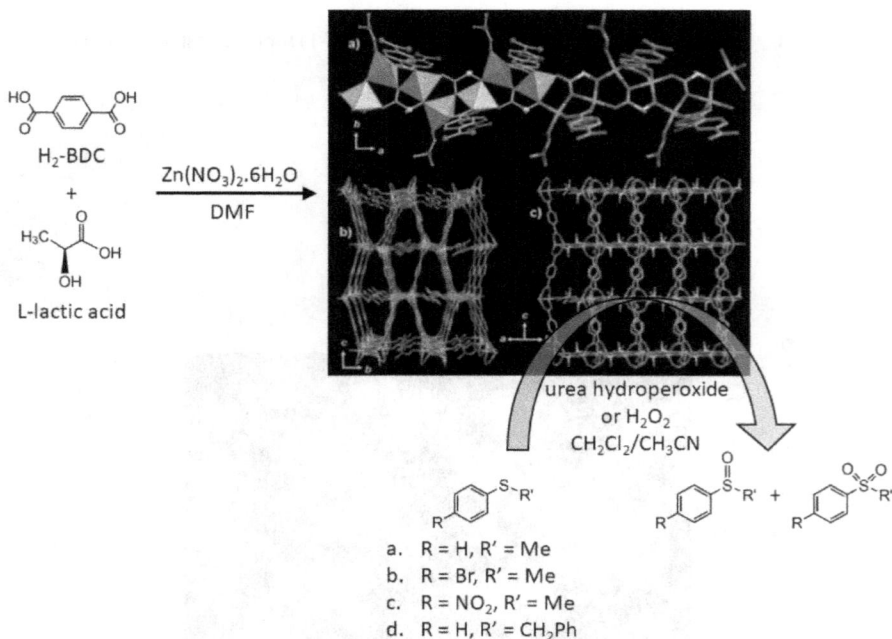

a. R = H, R′ = Me
b. R = Br, R′ = Me
c. R = NO₂, R′ = Me
d. R = H, R′ = CH₂Ph

FIGURE 6.3 Chemoselective oxidation of size-selective sulphide to sulphoxide by urea hydroperoxide or H_2O_2 with metal–organic framework [Zn_2(bdc)(l-lac)(dmf)].

(Reproduced after permission from reference 65.)

TABLE 6.2
Oxidation of Sulphides (a–d) Catalysed by [Zn$_2$(bdc)(l-lac)(dmf)]

Entry	Sulphide	Oxidant (equiv.)	Solvent	Conversion [%]	Selectivity [%]
1	A	UHP (2)	CH$_2$Cl$_2$	64	92
2	B	UHP (2)	CH$_2$Cl$_2$	58	83
3	c	UHP (2)	CH$_2$Cl$_2$	7	90
4	d	UHP (2)	CH$_2$Cl$_2$	3	Not measured
5	a	H$_2$O$_2$ 30% (3)	CH$_3$CN	92	100
6	a	H$_2$O$_2$ 90% (3)	CH$_2$Cl$_2$ + CH$_3$CN 10:1	100	87
7	b	H$_2$O$_2$ 90% (3)	CH$_2$Cl$_2$ + CH$_3$CN 10:1	85	100
8	b	H$_2$O$_2$ 90% (3)	CH$_2$Cl$_2$ + CH$_3$CN 10:1	58	100

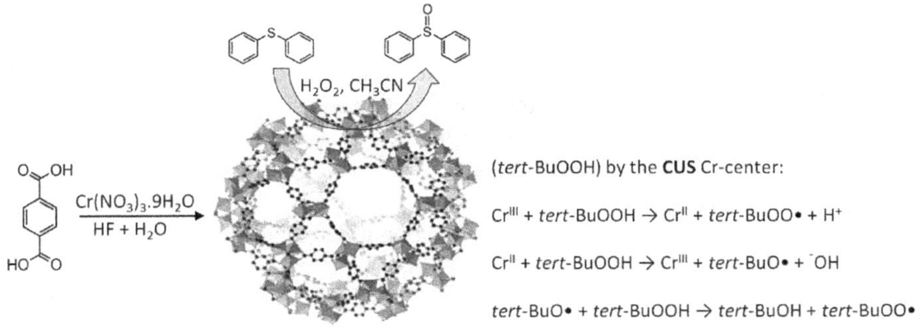

(*tert*-BuOOH) by the **CUS** Cr-center:

Cr^{III} + *tert*-BuOOH → Cr^{II} + *tert*-BuOO• + H$^+$

Cr^{II} + *tert*-BuOOH → Cr^{III} + *tert*-BuO• + ¯OH

tert-BuO• + *tert*-BuOOH → *tert*-BuOH + *tert*-BuOO•

FIGURE 6.4 Highly efficient catalytic oxidation of sulphide to sulphoxide with MOF, MIL-101.

(Reproduced after permission from reference 66.)

Size-selective catalytic oxidation at the node of MOF was explored in an excellent way by Fedin and co-workers [67]. They have attempted with linkers of different sizes, keeping functionality on linkers and metal ion constant. The pore size of the corresponding MOFs was dependent upon the size of the linkers used. This variation is reflected in the size-selective conversion and selectivity of sulphide to sulphoxide oxidation.

The previously reported MOF, [Zn$_2$-(bdc){(L)-lac}(dmf)] DMF, by the same author [65], synthesized by benzene-1,4-dicarboxylic acid (*bdc*), (*L*)-Lactic acid and Zn-metal ion, have pore size of 5 × 5 [Å]. Excellent conversion and selectivity were observed with small substrates like methyl phenyl thioether (PhSMe, entries 1 and 2 of Table 6.3). However, very poor conversion was observed for larger substrates like PhSCH$_2$Me (entry 4 of Table 6.2 and entry 3 of Table 6.3). When they have extended the frameworks with linkers of comparatively longer size, 4,4'-biphenyldicarboxylic

TABLE 6.3

Size-selective Sulphide Oxidation with 30% H_2O_2 over MOFs 1, 2 and 3

Entry	Sulphide	MOFs	[S]/[Zn]/H_2O_2	Conversion [%]	Selectivity [%]
1	PhSMe	3	3:1:9	92	100
2	PhSMe	3	10:1:15	94	98
3	PhSCH$_2$Ph	3	2:1:4	3	—
4	2-NaphSMe	1	12.5:1:25	78	99
5	PhSCH$_2$Ph	1	12.5:1:25	70	99
6	2-NaphSMe	2	12.5:1:25	57	98
7	PhSCH$_2$Ph	2	12.5:1:25	78	98.5

acid (*bpdc*) and 2,6-naphthalenedicarboxylic acid (*ndc*), the size of the pore of result-ing frameworks, [Zn$_2$-(bpdc){(*R*)-man}(dmf)]·2DMF (**2**) and [Zn$_2$-(ndc){(*R*)-man} (dmf)]·3DMF (**1**), increases respectively. As a result, although conversion with the substrate of larger size (entries 4–7 of Table 6.3) slowed down, selectivity observed was excellent.

Oxidation of unsaturated hydrocarbons, like cyclohexene, has been shown through Cu-based MOF [68] [Cu(H$_2$btec)(bipy)]$_\infty$ by Spodine and co-workers. Mixed ligand of 2,2'-bipyridine and 1,2,4,5-benzenetetracarboxylic acid was used for MOF synthesis through hydrothermal technique. In the single-crystal structure, Cu(II) centre has dis-torted square planar geometry. With *tert*-butyl hydroperoxide (TBHP), cyclohexene can be transformed into a mixture of the corresponding epoxide and 2-cyclohexenone, respectively. The reported MOF [Cu(H$_2$btec)(bipy)]$_\infty$, shows a high value for the con-version of cyclohexene (64.5%) at 75 °C for 24 h. The mechanism was proposed on the basis of literature report [69–71]. In the first step, distal O-atom of TBHP interacts with Cu(II) centre with an extension of the coordination number. Meanwhile, nucleo-philic attack of cyclohexene on the coordinated O-atom of TBHP followed by con-certed O-transfer leads to the formation of coordinated epoxide at the Cu-centre. With the subsequent elimination of epoxide, the catalyst was regenerated for the next round.

An interesting example of catalytic activity of cyanosilylation through Cu-based MOF was reported by Kaskel and co-workers [72]. The reported MOF [Cu$_3$(BTC)$_2$ (H$_2$O)$_3$·xH$_2$O], was synthesized using benzene-1,3,5-tricarboxylic acid and Cu(NO$_3$)$_2$·3H$_2$O by hydrothermal technique at 120 °C. At higher temperatures, the formation of Cu$_2$O (as by-product) along with the MOF was reported earlier. In the reported MOF, Cu$_2$-clusters are coordinated to carboxylate moieties to form a "paddle-wheel" unit (Figure 10, *top-left*), where O-atoms of carboxylate groups are found to be lying at the corner of a square. The Cu^{2+} ions are connected through a weak bond and the remaining site is occupied by a water molecule. Water molecules attached to Cu-centres were inward-facing (towards the pore of MOF, Figure 6.5, *bottom-right*). Thus, after removing coordinated solvent molecules, a catalytically active Lewis acidic site was on the interior of the pore wall. To approach the catalyti-cally active site, the substrate has to pass through the pore of MOF so, only substrate molecules of suitable size can pass through that pore of MOF. Hence, the catalytic process was size-selective.

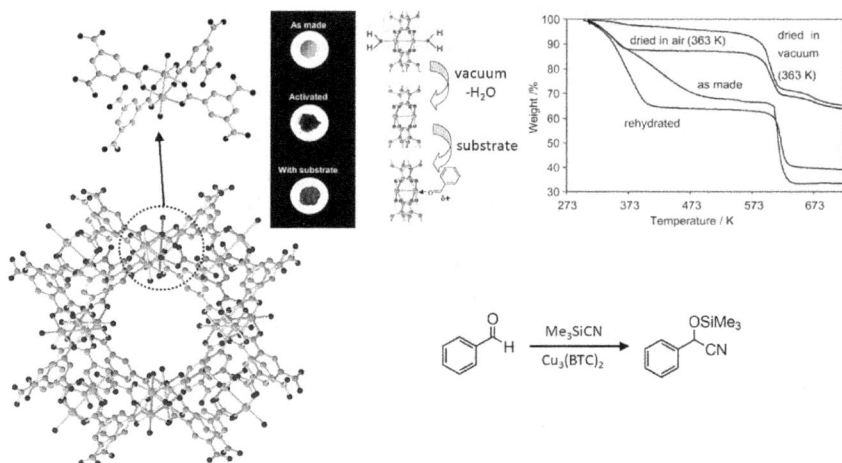

FIGURE 6.5 Cu-based MOF [Cu$_3$(BTC)$_2$(H$_2$O)$_3$·xH$_2$O] for cyanosilylation of benzaldehyde.

(Reproduced after permission from reference 72.)

Inspired from previous work, oxidation of aromatic hydrocarbon with Cu-based MOF in the presence of simple H$_2$O$_2$ is reported (Figure 6.6) by Baiker and co-workers [73]. In this example, CUS is created by applying heat/vacuum. With a dehydrated framework [Cu$_3$(BTC)$_2$], hydroxylation of aromatic hydrocarbon is performed in the presence of simple H$_2$O$_2$. Solvent-dependent selectivity was observed.

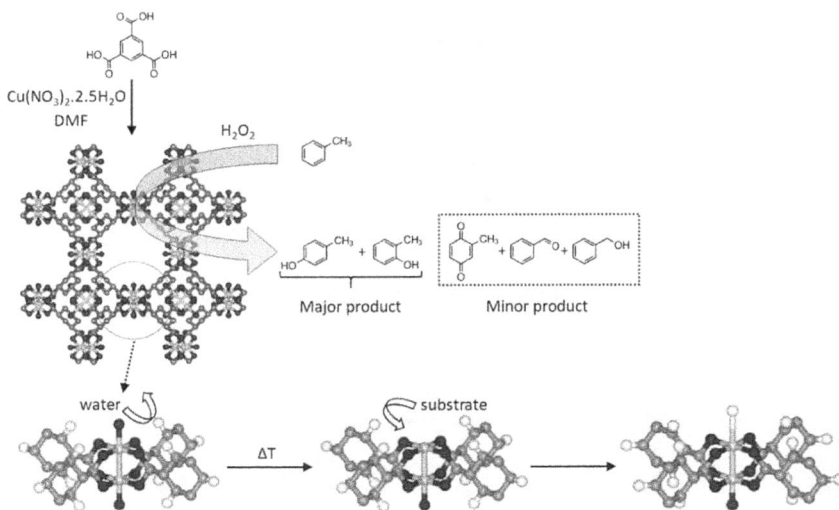

FIGURE 6.6 Hydroxylation of aromatic hydrocarbon at CUS of Cu-based framework [Cu$_3$(BTC)$_2$] with hydrogen peroxide.

(Reproduced after permission from reference 73.)

An interesting point about this work was that the original ligand benzene-1,3,5-tricarboxylate can be replaced with pyridine-3,5-dicarboxylate moieties. Analytical techniques such as X-ray-based techniques (powder XRD and XAS), thermal analysis and infrared spectroscopy showed that up to 50% of the original moiety (benzene-1,3,5-tricarboxylate) can be replaced with pyridine-3,5-dicarboxylate (Figure 6.7) without changing the framework topology. As expected, the strength of the binding site of the exchanged linker with Cu(II) will be changed which in turn, will change the selectivity of the product.

The order of reactivity observed in the order of benzene > toluene > p-xylene > o-xylene indicates that the size and steric hindrance of the reactants plays a crucial role. Thus, size-selective hydroxylation was possible with the reported MOF. In-depth studies show that the reaction is not proceeding through free radical mechanism and the proposed mechanism is shown in Figure 6.8.

CUS in the examples (discussed above) was created either by heating or by applying high vacuum. However, in most cases either the coordinative site is fully saturated or the framework is degraded by heating to remove coordinated solvent molecules. Rosseinsky and co-workers have reported [74] an elegant example where carboxylate moieties of Cu-carboxylate MOF get partially disconnected in the presence of proton. In other words, CUS in a MOF can also be created by acid treatment. It has been observed that both Lewis acid and Brønsted acid characters increased after protonation of MOF.

The presence of dual functionality, i.e., Lewis acid and Brønsted acid is further explored (Figure 6.9) in the conversion of glucose to 5-hydroxymethylfurfural (HMF) by Bao and co-workers [75]. Generally, HMF is produced from fructose by reaction with homogeneous acid, metal ions, acidic ionic liquid or zeolites as catalysts [76–79]. However, such conversion is not economical and it can be produced cheaply from glucose. The conversion of fructose to HMF requires dehydration of three molecules of water. The conversion of glucose to fructose has been reported either by solid-base catalyst [80] or by zeolites [81, 82]. Cr-based MOF, MIL-101(Cr),

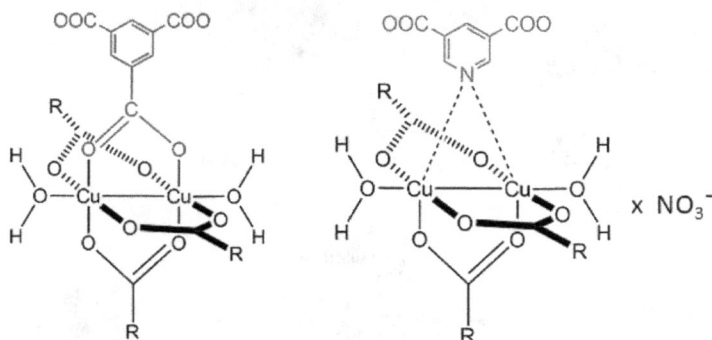

FIGURE 6.7 Pictorial representation of ligand (benzene-1,3,5-tricarboxylate) exchange with pyridine-3,5-dicarboxylate moiety.

(Reproduced after permission from reference 73.)

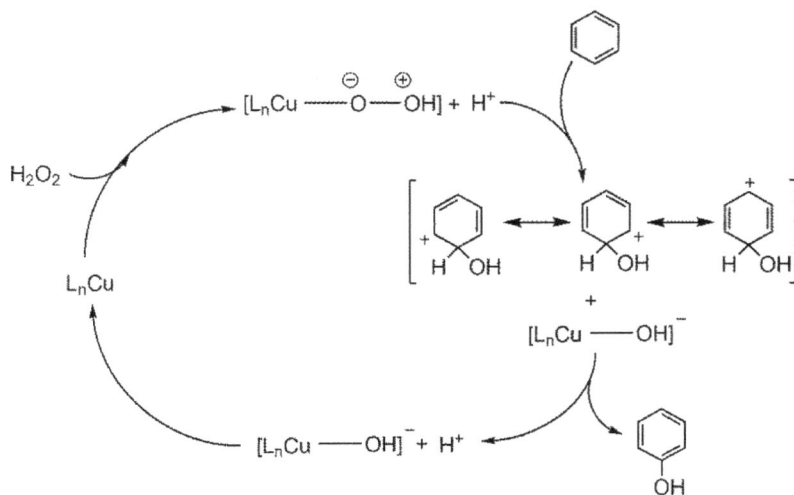

FIGURE 6.8 The proposed mechanism of oxidation of aromatic hydrocarbon with MOF $[Cu_3(BTC)_2]$ in the presence of H_2O_2.

(Reproduced after permission from reference 73.)

with $-SO_3H$ functional group on the linker has been explored for direct conversion of glucose to HMF. Lewis acidic site of MOF facilitates the conversion of glucose to fructose and $-SO_3H$ group present on the linker helps in the dehydration to eliminate three molecules of water to transform into HMF.

For asymmetric catalysis with MOFs, generally three approaches have been explored:

(i) Using chiral organic linker for MOF synthesis,
(ii) Using mixed organic linker for MOF synthesis, and
(iii) Post-synthetic modification on organic linker of MOFs (*vide infra*).

However, several unexpected problems were encountered using these approaches. For example, using a chiral organic linker for MOF synthesis would not be economical because of the multistep synthesis for the preparation of chiral ligand. Also, an unexpected interpenetration in the resulting MOF has been observed [83] that will limit the scope for heterogeneous catalysis. Using a mixed ligand/organic linker for MOF synthesis requires an extensive effort to get a particular MOF. As a novel approach, the coordinated solvent molecule has been replaced by chiral auxiliary units inside preassembled achiral frameworks, reported by Kim and co-workers (Figure 6.9) [60]. As discussed above, to get CUS, a metal of higher coordination number is preferred so that during the course of MOF synthesis, one or more coordination sites might be fulfilled by solvent molecule(s). Also, after removing coordinated solvent molecule(s), the resultant framework should be thermally stable. Kim and co-workers opted Cr-based MOF, MIL-101 [59] (discussed earlier in Figure 6.4), for this purpose

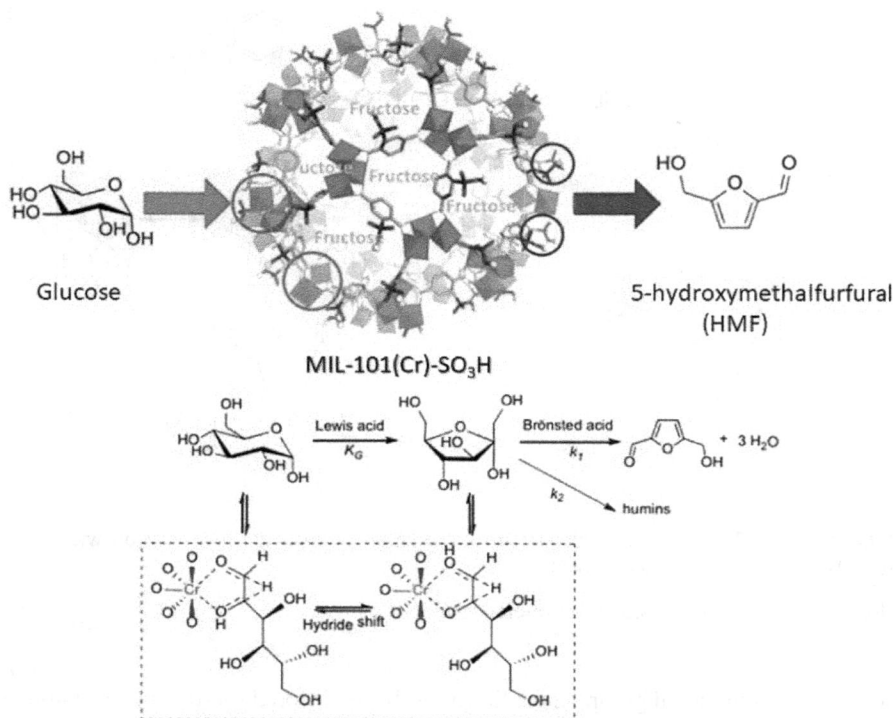

Glucose

MIL-101(Cr)-SO₃H

5-hydroxymethalfurfural (HMF)

FIGURE 6.9　Catalytic conversion of glucose to HMF by MOF [MIL-101(Cr)-SO₃H, *top*] bearing dual site: Lewis acidic site (*shown in large circle on the left*) and Brønsted acid (*shown in small circle on the right*) respectively. Chemical transformation along with possible mechanism is shown at the bottom.

(Reproduced after permission from reference 75.)

and CUS was filled by synthesized *L*-proline-based chiral ligands **L₁** and **L₂** to get the corresponding frameworks **CMIL-1** and **CMIL-2,** respectively (Figure 6.10). *L*-Proline and its derivatives are well-explored organocatalysts for enantioselective reactions, such as C–C bond-forming Aldol and Michael reactions [84]. Analytical analysis by powder X-ray diffraction (PXRD), thermogravimetric analysis (TGA), nitrogen adsorption measurements, IR spectroscopy and microanalysis of CMIL-1 and CMIL-2 suggest that the framework retains its architecture as that of parent MOF, MIL-101.

The catalytic performance of CMIL-1, CMIL-2 and their comparison with ligands **L₁** and **L₂** are summarized in Table 6.4. High conversion and enantioselectivity were observed with modified MOFs, CMIL-1 and CMIL-2. For example, when 4-nitrobenzaldehyde reacts with acetone (entry 1 of Table 6.4) in the presence of catalyst CMIL-1, 66% of yield and 69% of enantioselectivity were observed in neat condition (without any solvent). However, with ligand **L₁** as catalyst under homogeneous

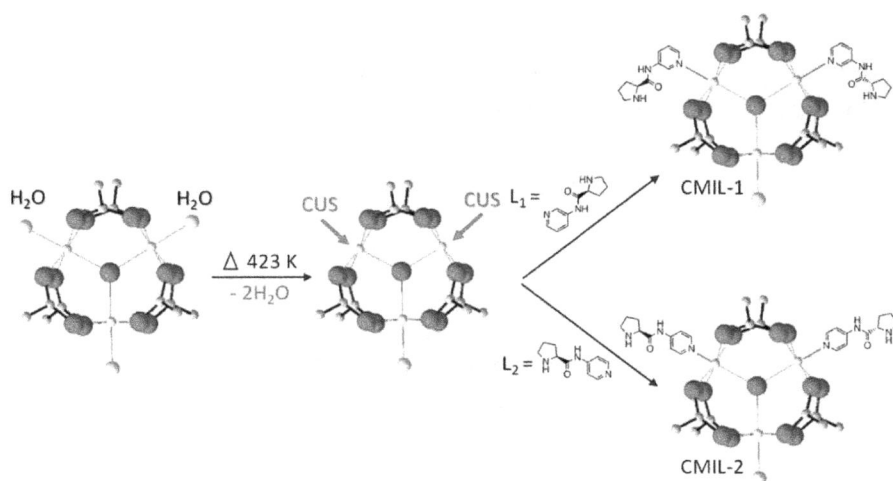

FIGURE 6.10 Schematic view of post-modification of **MIL-101** through *L*-proline-derived ligands at the coordinatively unsaturated chromium(III) centres.

(Reproduced after permission from reference 60.)

conditions, the yield was 58% but enantioselectivity was poor, only 29%. Better enantioselectivity with heterogeneous catalysts CMIL-1 and CMIL-2 were proposed as the restricted movement of the substrates in the confined environment along with multiple chiral inductions. Almost similar trends were observed by varying the substrate with MOFs CMIL-1 and CMIL-2 as catalysts. A slightly better performance in terms of enantioselectivity by CMIL-1 over CMIL-2 was proposed as bent shape of ligand L_1 (in CMIL-1) might impose additional restrictions to the substrate. Finally, the fact that the catalytic process was operational in the interior cavity of MOF and not on the surface was proved by using substrate **1e** (entry 8 of Table 6.4). A very poor yield in the presence of catalyst CMIL-1 and no detection of enantioselectivity were observed for substrate **1e**. The reason was that the size of the substrate was larger than the pore in the MOF, CMIL-1 so, the substrate was unable to pass through the pore, and therefore, no catalytic activity was observed in this case.

An interesting example of heterogeneous catalyst reported by Glorius, Kaskal and co-workers [85] where a novel design concept was used for MOFs synthesis and their pore-size estimation. Generally, the pore size of MOF is monitored by changing the length of the organic linker (Figure 6.11) [6, 86]. In the present report, the ligand used for MOF synthesis is the same but pore size depends upon the nature and polarity of the substituents (Figure 6.12) [87]. Two remarkable design concepts were used in the present report: (i) unlike previously reported examples, the chiral environment around active metal-centre is created by chiral groups/auxiliaries of the ligand, placed wisely in the close vicinity of the openly accessible metal node and (ii) instead of using metal of high coordination number to access CUS (by removing coordinated solvent molecule), an active site is used from the metal of multinuclear SBU.

TABLE 6.4

Catalytic Activities of L_1, L_2, CMIL-1 and CMIL-2 as Catalyst in Asymmetric Aldol Reaction between Different Aldehydes and Ketones

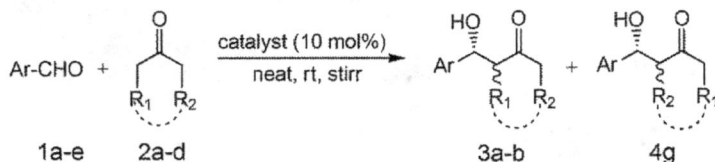

Entry		Ar	Substituents at ketone	Catalyst	Time (h)	Yield (%)	ee (%)
1	1a	4-Nitrobenzaldehyde	2a: $R_1=R_2=H$	CMIL-1	24	66	69
				L_1	24	58	29
				CMIL-2	24	59	63
				L_2	24	64	21
2	1b	Pyridine 4-carbaldehdye	2a	CMIL-1	16	91	76
				L_1	12	91	66
				CMIL-2	16	87	58
				L_2	16	89	37
3	1c	4-Chlorobenzaldehyde	2a	CMIL-1	40	74	70
				L_1	40	78	25
				CMIL-2	48	69	52
				L_2	48	75	23
4	1d	2-Napthaldehdye	2a	CMIL-1	60	80	63
				L_1	72	77	36
5	1a		2b: R_1, $R_2=-(CH_2)_3-$	CMIL-1	24	81	66
				L_1	24	76	49
6	1a		2c: R_1, $R_2=-CH_2-CH(t\text{-Bu})-CH_2-$	CMIL-1	36	86	68
				L_1	24	81	62
7	1a		2d: $R_1=H$, $R_2=CH_3$	CMIL-1	36	3g: 49	81
						4g: 27	69
8	1e	5-Formyl-1,3-phenylene bis(3,5-di-*tert*-butylbenzoate)	2a	CMIL-1	36	5	n.d.
				L_1	36	58	n.d.

Such active sites are reported to be found in M2 paddle-wheel units (M = Cu, Zn, Ni, Co, Mo) and can be used as Lewis acid catalysts [88].

The reported MOF $Zn_3(ChirBTB-1)_2$ and $Zn_3(ChirBTB-2)_2$ were synthesized by heating mixture solution of DFE (diethylformamide) with ligand **L1** ($H_3ChirBTB-1$) and **L2** $Zn_3(ChirBTB-2)_2$ and excess of $Zn(NO_3)_2 \cdot 4H_2O$. The pore size of MOF was estimated by absorption of dyes of different sizes (Figure 6.13). Generally, porosity of MOF is determined by gas physisorption for catalytic applications of fine chemical production, not the access of gases but rather pore access for large molecules in

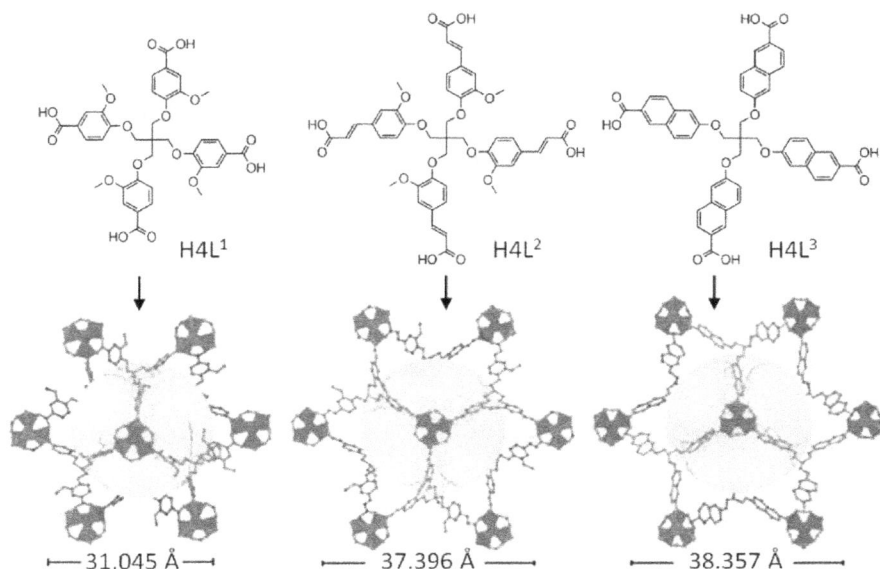

FIGURE 6.11 Tuning the pore size of MOF by increasing/changing the organic linker. (Reproduced after permission from reference 6.)

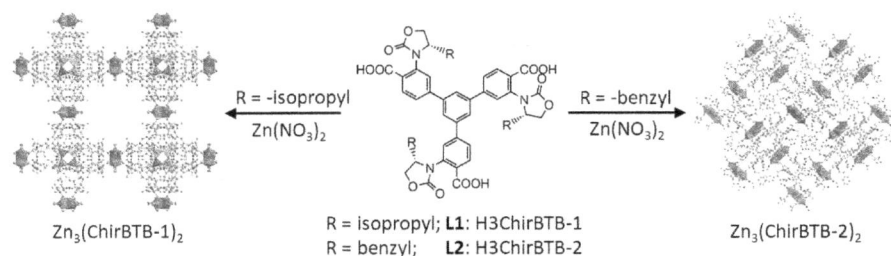

FIGURE 6.12 Nature and polarity of substituent (of ligand) dependent variation in pore size of MOF.

(Reproduced after permission from reference 87.)

solution is essential. Reichardt's dye was previously employed to probe the enormous pore size in MOF-177 [89]. In the present report, Fluorescein and Merocyanine dyes were absorbed by both $Zn_3(ChirBTB-1)_2$ and $Zn_3(ChirBTB-2)_2$. However, Reichardt's dye was absorbed (one molecule per unit) only by $Zn_3(ChirBTB-1)_2$. This observation clearly suggests that chiral group/auxiliaries of the ligand was responsible for different packing of both frameworks.

Since the pores of both MOFs were accessible for large molecules, both MOFs were tested for Mukaiyama aldol reaction of aromatic aldehydes with 1-methoxy-1-(trimethylsiloxy)-2-methyl-1-propene. Such a reaction requires a strong Lewis acid

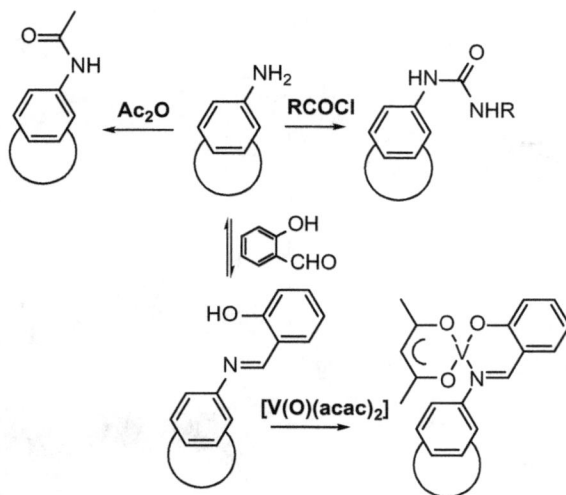

FIGURE 6.13   Modification on linker surface through post-functionalisation method.

catalyst for chemical transformation. Mukaiyama aldol reaction was tested through MOF as catalyst first by Long and co-workers [37] and later by Cohen et al [90]. through post-synthetic approach. However, enantioselectivity in Mukaiyama aldol reaction was not reported earlier with MOF as catalyst. The results obtained in the present report are summarized in Table 6.5. Two different solvent systems were checked with the assumption to replace the coordinated solvent with catalytically

TABLE 6.5

Results of Mukaiyama Aldol Reactions using Zn_3(ChirBTB-1)$_2$ and Zn_3 (ChirBTB-2)$_2$ as Catalyst in Two Different Solvent Systems

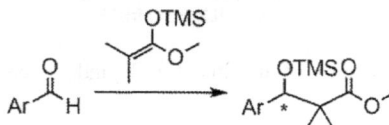

Ar	Catalyst	Yield (%)	ee (%)
Solvent: *Dichloromethane*			
Ph	Zn_3(ChirBTB-1)$_2$	83	0
1-Naph	Zn_3(ChirBTB-1)$_2$	31	40
Ph	Zn_3(ChirBTB-2)$_2$	66	8
1-Naph	Zn_3(ChirBTB-2)$_2$	0	—
Solvent: *n-Heptane*			
Ph	Zn_3(ChirBTB-1)$_2$	77	9
1-Naph	Zn_3(ChirBTB-1)$_2$	77	16
Ph	Zn_3(ChirBTB-2)$_2$	43	6

active site because removal of coordinated solvent was not possible with high vacuum without collapsing of the framework.

6.5.2 Surface Modification or Functionalisation of Linker Unit

Surface modification or surface engineering [40, 91] of the linker unit is another approach for the synthesis of a heterogeneous catalyst. In MOFs, where an organic linker unit without any functional group is used, the guest molecule interacts with weak dispersive force only. However, if the guest molecule(s) interact either with hydrogen bond or through coordinate bond then the interaction between the guest and the wall of MOF will be strong. As a result, the transition state can be stabilized much better in a confined environment and therefore selection of the product will be profound.

As discussed above, in order to create a CUS, a metal of higher coordination number is preferred.

However, it would not be economical to choose lanthanides for a node with CUS. Moreover, when CUS is introduced by heating or by applying high vacuum, the framework occasionally collapses. To remove this drawback, mainly three approaches are used called as 'post synthetic modification'.

(i) *Post-synthetic activation of functional group*: Functional groups like amine, amide, alcohol, urea, thiourea, etc., are able to bind with active group of substrates through hydrogen bonding. Such bonding activates substrate molecules to undergo the catalytic process. If organic linkers are decorated with such functionalities, then the resulting MOF would be able to facilitate the catalytic process in its confined environment through hydrogen bonding. During the synthesis of MOF, a protected form of functional group is used and deprotected once MOF is synthesized.

(ii) *Post-functionalisation of linker unit*: In this approach, the catalytic unit is tagged after MOF formation with organic linker bearing a suitable functional group (Figure 6.13). Such an approach provides ample opportunity to tag a variety of catalytic sites, for example, if the amino group is present on the organic linker of MOF, it may be modified either to amide, urea or thiourea functionality for guest binding through hydrogen bond or metal as the catalytic site may be introduced through Schiff's base formation. Thus, the pore of MOF will allow size-selective guest molecules to approach the catalytic site.

(iii) *Post-metalation of linker unit*: In this approach, a linker unit bearing an additional functional group, which does not interfere/interrupt the metalation process during MOF synthesis, is chosen. Once the MOF is synthesized then it is directly charged (called as post-synthetic modification) again with another metal with which we want to carry out catalysis. During post-metalation, the MOF is synthesized with metal ion and post-modification is done with another metal ion. It is worthy to mention that both metal ions are doing their jobs: one metal ion provides porous architecture, giving strength and support to framework whereas the catalytic process is carried out with another metal ion. Some interesting and selective examples will be discussed working through this approach.

A homochiral porous MOF for heterogeneous asymmetric catalysis was reported by Lin and co-workers [92] (Figure 6.14). Ligand (R)-7,7′-dichloro-2,2′-dihydroxy-1,1′-binaphthyl-4,4′-bipyridine possesses bipyridyl as primary functionality and orthogonal chiral 2,2′-dihydroxy group as secondary functionality. Homochiral MOF [Cd$_3$Cl$_6$L$_3$], 4DMF, 6MeOH, 3H$_2$O was synthesized at room temperature by diffusion of Et$_2$O into DMF/MeOH mixture solution of ligand and CdCl$_2$. The resultant MOF was charged with Ti(O^{i}Pr)$_4$ that binds with chiral hydroxyl group of the organic linker to afford active catalyst. The reported catalyst performs asymmetric catalysis with activity and stereoselectivity rivaling its homogeneous counterparts.

In the above reported example, the author has studied catalytic activity with substrates of varying size and compared the study with BINOL/Ti(O^{i}Pr)$_4$. The conversion is gradually decreasing with substrates 4′-G$_1$′OPh, 4′-G$_1$OPh (entries 6 and 7 of Table 6.6). The author has proposed that hinderance arises (as size of substrate increases) in order to pass through the pore of MOF and as a result conversion decreases. The size of substrate 4′-G$_2$′OPh (entry 8 of Table 6.6) is large enough for it to not able to pass through the pore of MOF·Ti, and thus, no conversion observed

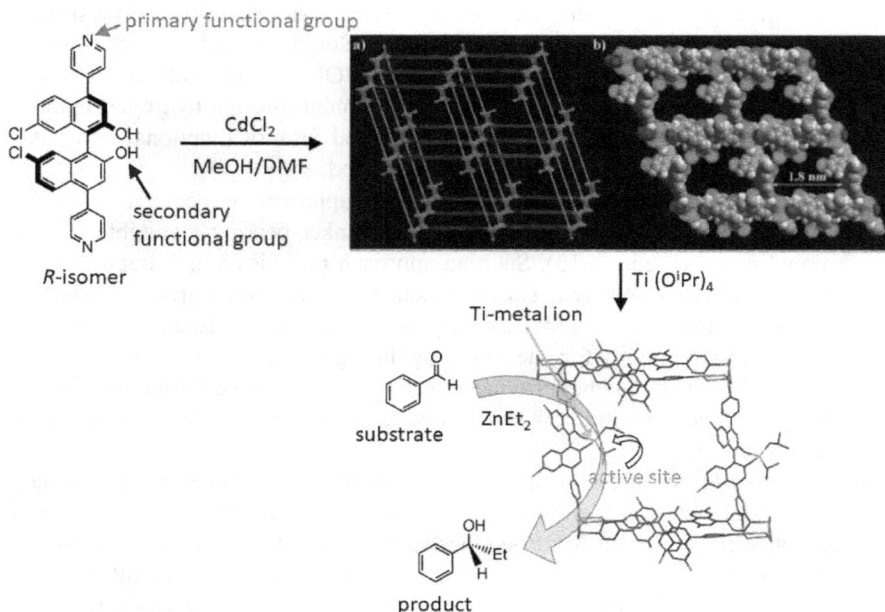

FIGURE 6.14 Binaphthyl moiety-based homochiral porous MOF for heterogeneous asymmetric catalysis was reported by Lin and co-workers.

(Reproduced after permission from reference 92.)

TABLE 6.6
Ti(IV)-Catalysed $ZnEt_2$ Additions to Aromatic Aldehydes

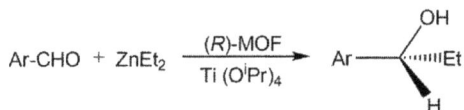

$$Ar\text{-}CHO + ZnEt_2 \xrightarrow[Ti(O^iPr)_4]{(R)\text{-}MOF} Ar\text{—}\overset{OH}{\underset{H}{\underset{|}{C}}}\text{''''}Et$$

		BINOL/Ti(O^{i}Pr)$_4$		MOF·Ti	
		Conversion		Conversion	
Entry	Substrate	(%)	ee (%)	(%)	ee (%)
1	1-Napthaldehyde	>99	94	>99	93
2	Benzaldehdye	>99	88	>99	83
3	4-Cholorbenzaldehdye	>99	86	>99	80
4	3-Bromobenzaldehdye	>99	84	>99	80
5	4'-G$_0$OPh	>99	80	>99	88
6	4'-G$_1$'OPh	>99	75	73	77
7	4'-G$_1$OPh	>99	78	63	81
8	4'-G$_2$'OPh	95	67	0	—
	R=	–CH$_3$			
Dentritic aldehyde	Dendron	G$_0$	G$_1$'	G$_1$	G$_2$'
	Est. size	0.8 nm	1.45 nm	1.55 nm	2.0 nm

for this substrate. Thus, size-selective asymmetric alkylation of aromatic aldehydes is reported with excellent selectivity.

Lin and co-workers also reported network-structure-dependent catalytic activity [93] by MOFs [Cd$_3$L$_4$(NO$_3$)$_6$]·7MeOH·5H$_2$O (**MOF-1**) and [CdL$_2$(H$_2$O)$_2$][ClO$_4$]$_2$·DMF·4MeOH·3H$_2$O (**MOF-2**). Ligand (R)-6,6'-dichloro-2,2'-dihydroxy-1,1'-binaphthyl-4,4'-bipyridine (**L**) was charged with Cd(NO$_3$)$_2$·4H$_2$O and Cd(ClO$_4$)$_2$·6H$_2$O to afford **MOF-1** and **MOF-2**, respectively (Figure 6.15). CO$_2$ adsorption studies suggest that evacuated sample of **MOF-1** and **MOF-2** have specific surface area of 772.3 m^2 g^{-1} and 370 m^2 g^{-1} and pore volume of 0.25 mL g^{-1} and 0.16 mL g^{-1}, respectively.

An attempt was made to generate heterogeneous asymmetric catalysts by activating Lewis acidic metal centres (of Ti(O^{i}Pr)$_4$) with the chiral dihydroxy groups that are present as the orthogonal secondary functionalities in the porous **MOFs 1** and **2**. By treatment of **MOF-1** with an excess of Ti(O^{i}Pr)$_4$ in toluene afforded active catalyst **MOF-1·Ti**. The resultant active catalyst (**MOF-1·Ti**) successfully transforms

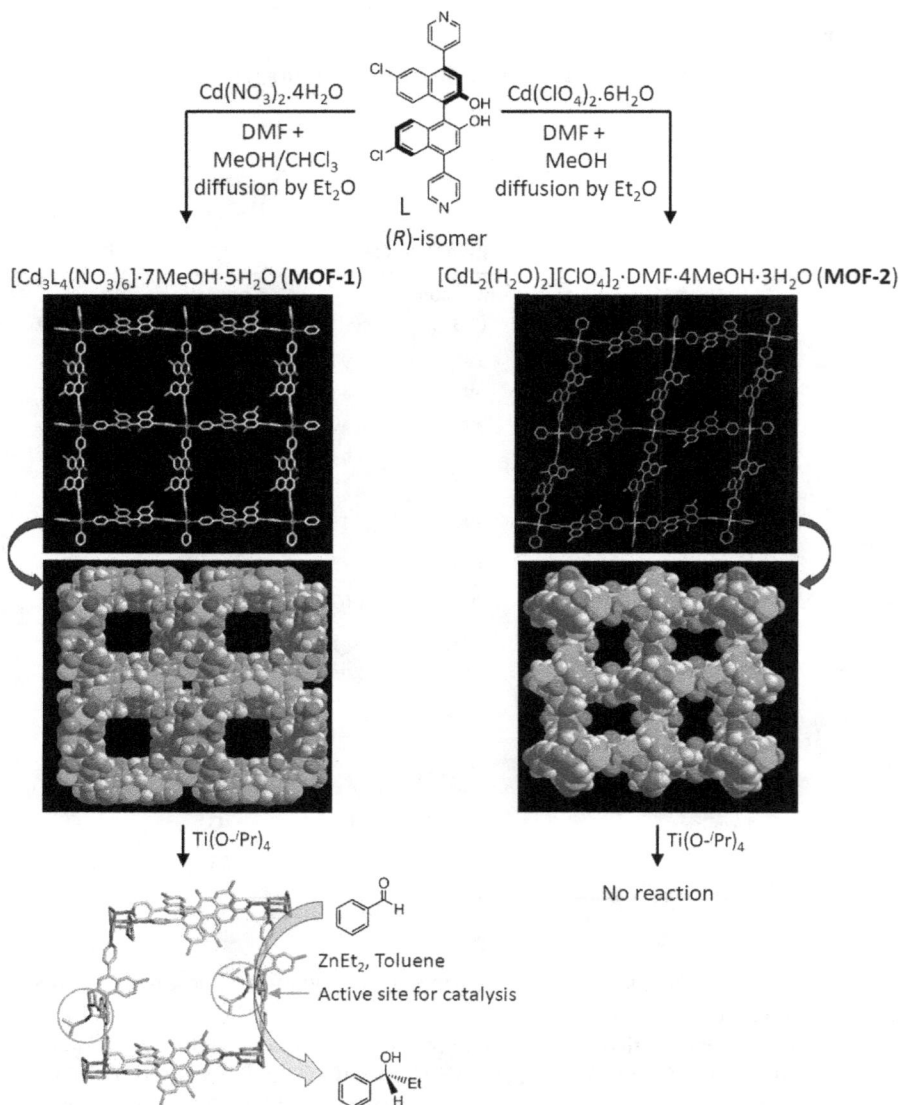

FIGURE 6.15 Network-structure-dependent heterogeneous catalysis of alkylation of aromatic aldehydes.

(Reproduced after permission from reference 93.)

1-naphthaldehyde to the corresponding secondary alcohol with 90% enantiomeric excess (ee). Slightly better selectivity was reported with benzaldehyde (entry 4 of Table 6.7) with catalyst (**MOF-1·Ti**) than the original ligand (L, entry 5).

However, such a transformation was not observed with an activated catalyst of **MOF-2** with Ti(O^{i}Pr)$_4$. The author explains this discrimination through a single

TABLE 6.7
Addition of Diethylzinc to Aromatic Aldehyde Catalysed by MOF-1·Ti

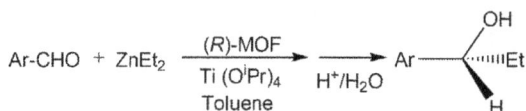

$$Ar\text{-}CHO + ZnEt_2 \xrightarrow[\substack{Ti\,(O^iPr)_4 \\ Toluene}]{(R)\text{-}MOF} \xrightarrow{H^+/H_2O} Ar\text{—}\overset{OH}{\underset{H}{\text{C}}}\text{\tiny{········Et}}$$

Entry	Ar	Catalyst	Loading (%)	Conversion (%)	ee (%)
1	1-Naphthyl	MOF-1·Ti	12	>99	90.0
2	4-CH$_3$Ph	MOF-1·Ti	12	>99	84.2
3	4-CH$_3$Ph	MOF-1·Ti	25	>99	84.9
4	Ph	MOF-1·Ti	12	>99	81.9
5	Ph	L	20	>99	78.6
6	3-BrPh	MOF-1·Ti	12	>99	71.0
7	4-ClPh	MOF-1·Ti	12	>99	60.2
8	4-CF$_3$Ph	MOF-1·Ti	12	>99	45.0

crystal structure of **MOF-1** and **MOF-2**. The crystal structure of **MOF-2** reveals that the pyridyl and naphthyl rings are mutually perpendicular to each other. Moreover, chiral hydroxyl functional group was almost perpendicular to each other and this may be the reason that no complexation was possible when **MOF-2** in toluene was charged with an excess of Ti(O^{i}Pr)$_4$.

Stability is a prominent factor in ensuring better performance of a heterogeneous catalyst. Thermal stability of MOF can be increased either by using mixed ligand or by using "paddle-wheel" as SBU. Thermal stability of MOF also increases by the interpenetration of the network. Apart from network stability, the catalytic site also must be stable for obtaining better output. Jacobsen's catalyst, used for the epoxidation process is very sensitive and its performance depends upon several factors [94]. Loss of catalytic activity (with time) of Jacobsen's catalyst typically, is associated with oxidation of the salen ligand. If salen oxidation is facilitated by reactive encounters with other catalyst molecules, immobilisation of the catalytic site should prevent encounters. It will also extend the catalyst lifetime. An example for improvement of epoxidation process was reported by Hupp, Nguyen and co-workers [95] with Mn-salen based MOF and biphenylene dicarboxylate as secondary ligand. Hupp, Nguyen and co-workers have attempted to sort out this drawback by immobilisation of salen unit as part of the framework.

Porous framework Zn$_2$(bpdc)$_2$L·10DMF·8H$_2$O, was synthesized through solvothermal technique by heating DMF mixture solution of (*R,R*)-(−)-1,2-cyclohexanediamino-*N,N*′-bis(3-*tert*-butyl-5-(4-pyridyl)-salicylidene)Mn^{III}Cl as ligand **L** and biphenylene dicarboxylic acid as secondary ligand with Zn(NO$_3$)$_2$·6H$_2$O.

The catalytic activity of the reported MOF for asymmetric epoxidation was examined with 2,2-dimethyl-2*H*-chromene as substrate and 2-(*tert*-butylsulphonyl)iodosylbenzene used as oxidant. The author has compared the catalytic activity of MOF with their building block, i.e., ligand **L** [(*R,R*)-(−)-1,2-cyclohexanediamino-*N,N*′-bis(3-*tert*-butyl-5-(4-pyridyl)salicylidene)-Mn^{III}Cl]. Minor decrease in selectivity with MOF was observed in comparison to ligand **L** (82% ee for MOF vs. 88% ee for free **L**). The fact that a decrease in enantioselectivity is observed when electron-withdrawing group is introduced, is well explained in reported literature [96]. The author has also explained that in the reported MOF, the electronic nature of ligand **L** will be changed when pyridyl-moiety of the ligand will bind with Zn-cation.

The author has reported that enantioselectivity was constant even after the third cycle of catalysis. However, a minor decrease in TON was observed. It was explained that recycling was accompanied by MOF particle fragmentation. Inductively coupled plasma (ICP) spectroscopy of the product solution, after removal of MOF particles showed that roughly 4–7% of the manganese initially present in the framework material was lost per cycle, either as molecular species or as particles too small to be removed by filtration through Celite. It has been observed that free manganese, present in the solution did not catalyse the reaction. Thus, life-time of Jacobsen's type catalyst can be increased by their immobilisation, in the form of MOF.

Another example of heterogeneous catalyst for acyl-transfer was reported again by Hupp, Nguyen and co-workers [97] using metalloporphyrin as strut and 1,2,4,5-tetrakis(4-carboxyphenyl)-benzene as bridging unit (Figure 6.16).

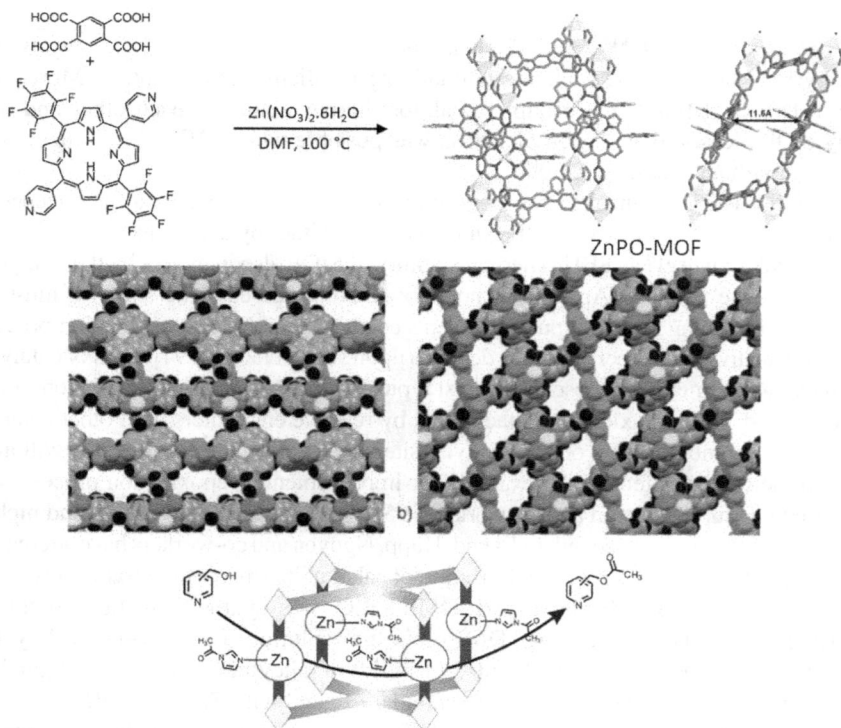

FIGURE 6.16 Metalloporphyrin as strut of MOF (*top*), for heterogeneous catalysis of acyl-transfer. Uniform distribution of well-defined pore (*middle*) to allow guest of suitable size. Schematic representation of catalytic activity of reported ZnPO-MOF (*bottom*).

(Reproduced after permission from reference 101f.)

Metalloporphyrin is well explored as a molecular catalyst in an artificial system or as an active site of metalloenzymes. However, to use metalloporphyrin as a catalytically active site, one has to cross through three major design challenges: (i) porphyrinic MOFs possess large open pores that are unusually susceptible to collapse upon removal of solvent(s); (ii) it is very difficult and challenging to prevent porphyrin metal sites from doubling as nodes for strut coordination, thereby blocking potential catalytically active sites; (iii) attempts to incorporate free-base porphyrins as struts (which would then be available for post-synthesis metalation) are frustrated by the tendency of the porphyrin ligand to scavenge and coordinate metal ions present in the initial MOF synthesis.

Mixed ligand strategy and paddle-wheel as SBU were used to empower thermal stability to the framework. Porous framework named as **ZnPO-MOF**, was synthesized under solvothermal condition by using (5,15-dipyridyl-10,20-bis (pentafluoro-phenyl))porphyrin and 1,2,4,5-tetrakis(4-carboxyphenyl)-benzene with $Zn(NO_3)_2 \cdot 6H_2O$. The unique feature of this reported MOF is the non-interpenetration of network. The paddle-wheel unit as SBU generally shows interpenetration of framework [67, 95, 98]. **ZnPO-MOF** shows catalytic acyl-transfer between 3-pyridylcarbinol (3-PC) and *N*-acetylimidazole (NAI) by about 2420 times faster than uncatalysed reaction. No significant differences in the rate of reaction were observed among the three isomers of PC. This observation was in stark contrast to that observed with discrete supramolecular catalysts [99, 100], which catalyse the transfer of an acyl group to the 3-PC and 4-PC much more effectively than to 2-PC. The proposed mechanism of acyl-transfer is shown in Figure 6.17.

6.5.3 CENTRAL PORE/CAVITY OF MOFs

The confined environment of MOF has also been explored for heterogenous catalysis. Some homogeneous catalysts like metalloporphyrins have been well-explored in literature, typically, for hydroxylation and epoxidation reactions [101]. However, the formation of bridged μ-oxide dimer in metalloporphyrin creates an obstacle for the access of the catalytically active site, and oxidative self-degradation [102] limits its life-time for wide applications. In order to remove this drawback, immobilisation of metalloporphyrin in solid matrices would be a worthful approach [103–106]. Porous inorganic solids (i.e., zeolites X and Y [107], mesoporous silicates [108],

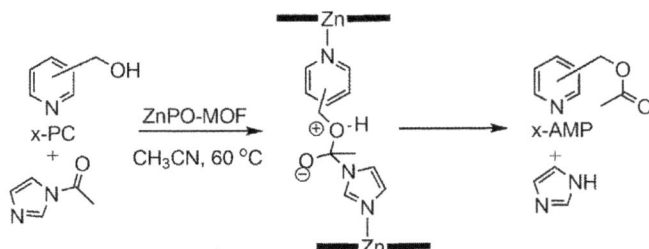

FIGURE 6.17   Possible proposed mechanism of acyl-transfer by **ZnPO-MOF**.

(Reproduced after permission from reference 101f.)

silica surfaces [109]) have been explored as solid-state matrices for immobilisation of metalloporphyrins. However, these systems also are faced with limitations such as aggregation of the catalyst molecules and limited catalyst loadings resulting in potential non-periodic distribution, and/or leaching of the catalyst.

Encapsulation of such catalysts in systems like MOF is a tough and highly challenging task (Figure 6.18). For better performance of such encapsulated catalysts, the host system should have the following characteristic features: (i) large cavities, suitable for the encapsulation of one guest porphyrin molecule per cavity, with relatively reduced openings that prevent the catalyst leaching while allowing diffusion of reactants and products; (ii) mild synthesis conditions and the presence of framework-porphyrin directing interactions, for example, electrostatics, permitting the one-step framework construction and encapsulation of the free-base porphyrin; (iii) maintained framework integrity upon post-synthesis metalation of the encapsulated porphyrin (e.g., framework stability in aqueous media) and under the investigated catalytic oxidation conditions; (iv) low affinity for the oxidation products, thus allowing for their diffusion into the bulk solution and ease of separation (e.g., filtration) from the heterogeneous catalyst.

Eddaoudi and co-workers have reported [36] an interesting example of MOF possessing most of the features, as discussed above, to encapsulate sensitive catalysts like metalloporphyrin. The MOF formed, using 4,5-imidazole dicarboxylic acid (H3ImDC) and $In(NO_3)_3$ was shown to be an unusually robust MOF (named rho-ZMOF), anionic in nature. The author has wisely used the cationic form of porphyrin to facilitate the encapsulation process with anionic MOF (as host) through charge

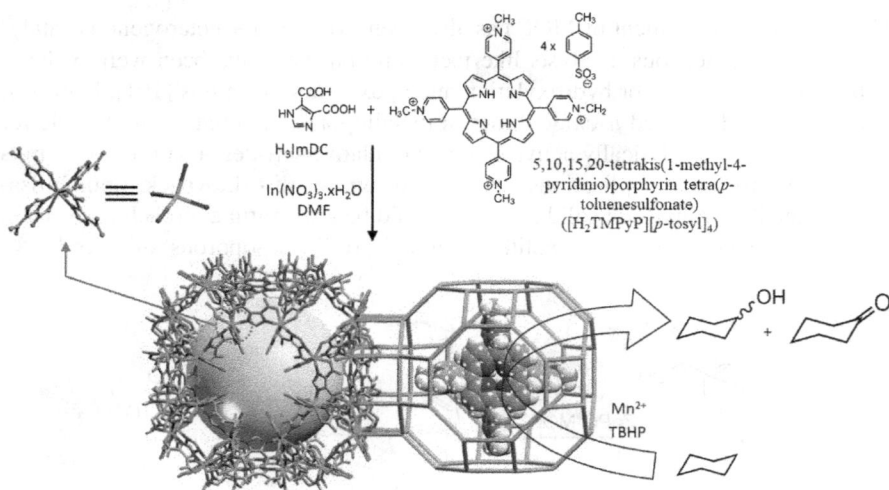

FIGURE 6.18 Solvothermal synthesis of cationic porphyrin (as guest) encapsulated MOF (as catalyst) for oxidation of hydrocarbon.

(Reproduced after permission from reference 36.)

interactions. The author has reported cationic porphyrin encapsulated MOF synthesis by solvothermal technique, by heating N,N'-dimethylformamide (DMF)/acetonitrile (CH_3CN) mixture solution of $In(NO_3)_3 \cdot xH_2O$ and 4,5-imidazoledicarboxylic acid (H3ImDC) in the presence of 5,10,15,20-tetrakis(1-methyl-4-pyridinio)porphyrin tetra(p-toluenesulphonate) ([H2TMPyP] [p-tosyl]$_4$) to afford dark red cubic-like crystals. Through various analytical techniques, encapsulation of cationic porphyrin, as guest was confirmed. The author successfully performed metalation of guest porphyrin with several transition metal ions and with their manganese (Mn) derivative catalytic activity for oxidation of hydrocarbon was tested and the results were compared with previously reported catalysis by zeolite-encapsulated Fe-porphyrin [111b, 111c, 109] (summarized in Table 6.8). Unlike previously reported examples, simple *tert*-butylhydroperoxide was used as oxidant instead of iodosobenzene. Cyclohexanol and cyclohexanone were the products of the reaction and no side product formation was observed. The observation made from Table 6.8, clearly suggests that almost in all respects like TON, yield of reaction, catalyst loading, the performance of present catalyst was simply outstanding although slightly longer time duration was required for completion of the reaction.

The encapsulation of reactive catalyst is a fantastic and efficient approach provided during their formation the catalyst should not interfere with the host and also pore of the host should be of suitable size to give a confined environment to the reactive catalyst and the catalyst should not squeeze out of the host during the catalytic process.

Polyoxometalates (POMs) anions are another example of an excellent catalyst widely used for variety of catalytic applications. Its excellent redox properties, high acidity, good thermal as well as chemical stabilities drag scientists' attention towards catalysis. The necessary requirement is to disperse on solid support because of their very low surface areas. Agglomeration, non-uniform distribution, leaching of POM on traditional supports like activated carbon, silicas are drawbacks for its practical applications. To resolve this issue, the first example reported by Férey and co-workers is to use *meso*-MOFs like MIL-101 as support to load POM anions [59]. Recently, encapsulation of POM anions inside *meso*-MOF as host has been reported [110] in MIL-100(Fe).

Generally, the POMs are encapsulated by one-pot *in situ* synthesis method. For example, POM@MIL-100(Fe) has been synthesized [111] by heating the mixture of $Fe(NO_3)_3 \cdot 9H_2O$, H3BTC, HPW and deionized water in one reactor (Figure 6.19). This encapsulated catalyst is highly efficient for esterification and acetalisation reactions.

Among POM anions with Keggin structure $[XM_{12}O_{40}]^{n-}$ (where X = P, Si, Al, etc. and M = W, Mo), phosphotungstic acid (PTA) is the strongest heteropoly acid and has good thermal stability [112]. PTA acts as a strong Brønsted acid. Recently, encapsulated PTA has been explored as a heterogeneous catalyst for the transformation of phenol to 4,4'-bisphenol with formaldehyde [113]. The proposed mechanism is shown in Figure 6.20. In the reported example, encapsulated PTA acts as proton donor–acceptor.

In catalysis, the surface area of the catalyst also matters a lot for sufficient and efficient catalysis. For such catalysts, solid support is required for homogeneous

TABLE 6.8
Summary of Cyclohexane Oxidation Reactions Heterogeneously Catalysed by Encapsulated Metalloporphyrins in Solid Matrices (MOF)

Reference	Catalyst	Cyclo-hexane	Temperature, time	Oxidant	TON[a]	Yield[b] (%)	Loading[c] (%)
[111c]	Fe(III)-P[d] encapsulated in zeolite Y. Used 0.25 µmol catalyst. (1 porphyrin ring per 40 available super cages)	14.6 mmole	r.t., 6 h	PhIO[e] 5 µmol	7.6	38	5
[111b]	Fe(III)-TMPyP[f] encapsulated in zeolite X. Used 0.055 µmol catalyst. (1 g solid = 1.1 µmol catalyst)	2.66 mmole	r.t., 7.5 h	PhIO 1 µmol	10	60	5.5
[113]	Fe(III)-TMPyP supported on silica surface and matrices. Used 0.25 µmol catalyst	200 µL	r.t., not reported	PhIO 25 µmol	2–20[g]	2–20	1
Present report	Mn-TMPyP encapsulated in rho-ZMOF. Used 2.9 µmol catalyst. (30 mg solid = 1 µmol catalyst)	neat	65 °C, 24 h	TBHP[h] 77 µmol	24	91.5	3.8

[a] TON = (moles of cyclohexanol + 2X moles of cyclohexanone)/moles of catalyst.
[b] Based on moles of oxidant.
[c] Moles of catalyst/moles of oxidant.
[d] P = tris(4-N-methyl-pyridyl)-mono(pentafluorophenyl) porphyrin.
[e] PhIO = iodosobenzene
[f] TMPyP = 5,10,15,20-tetrakis(1-methyl-4-pyridinio)porphyrin.
[g] Values reported depend on the different catalyst systems used in reference 3.
[h] TBHP = $tert$-butylhydroperoxide.

distribution of catalysts like palladium over activated charcoal for hydrogenation process. MOFs facilitate additional advantages like leaching of catalyst in order to avoid product contamination with metal ion.

The pores/channels of MOF provide a confined space for nucleation of the metal-nanoparticles (MNPs). The functionalised groups dictate the size and dispersion of

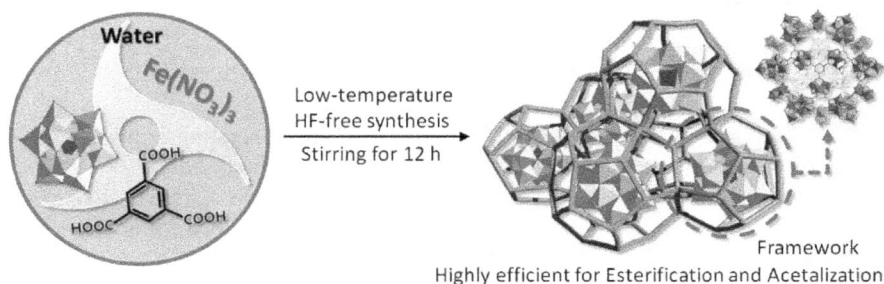

FIGURE 6.19 In situ incorporation of POM into the frameworks of MOFs for catalysis of esterification and acetalisation reactions.

(Reproduced after permission from reference 111.)

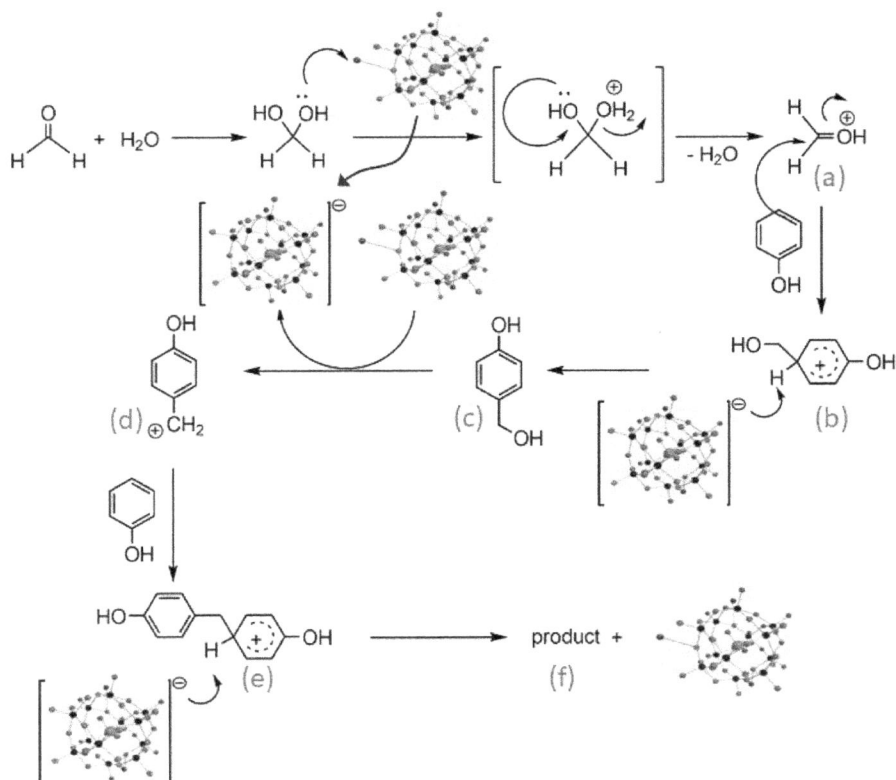

FIGURE 6.20 Proposed plausible mechanism for hydroxyalkylation of phenol and formaldehyde to bisphenol (based on 4,4′-isomer). The framework of PTA catalyst is ball-and-stick.

(Reproduced after permission from reference 113.)

MNPs. Various techniques used for such purposes like: (i) chemical vapor deposition [114–116], (ii) solution infiltration [117–119], (iii) solid grinding [20, 120, 121], etc. These methods are used to encapsulate or impregnate MNPs within MOF pores/ channels to prepare MNPs@MOFs [122].

Zhao, Yuan and co-workers recently reported [123] an interesting example of site-selective reduction of α,β-unsaturated aldehyde to α,β-unsaturated alcohol through MOF-based catalysts [MIL-101(Cr)@Pt@MIL-101(Fe)] with sandwich-like core–shell configuration, in which the MNPs (Pt or Ru) are located between an inner core and an outer shell composed of MIL-101 with metal nodes of Fe^{3+}, Cr^{3+} or both. α,β-Unsaturated alcohol formation is one of the interesting topic for synthetic chemists because of its huge utility, especially in perfumes and pharmaceutical industry. However, hydrogenation of C=C bond is thermodynamically favoured over C=O bond [124].

Therefore, developing an effective catalyst for site-selective hydrogenation of α,β-unsaturated aldehyde is a tough and challenging task [125–130].

There are two more strategies [131] of heterogeneous catalysis: (i) replacing metal at the node with another metal and (ii) replacing organic linker of MOF with another linker. Although these two strategies facilitate to form resultant MOF of our desire, distribution of such changes is found to be not equally distributed throughout the framework. Their modifications are in progress and new findings are coming out with time.

In conclusion, MOFs should have the following characteristics in order to be used as heterogeneous catalysts: (i) it should possess suitable pore-size, should be easily accessible and if required can be programmed accordingly, (ii) should be chemically and thermally stable during the course of reaction, (iii) should be reusable, and most importantly, (iv) it should be overall cost-effective. Encapsulation of reactive catalyst or their fixation as an organic linker is an interesting approach not only to provide stability or life-time to that catalyst but also to increase the yield and efficient selectivity of the product. Although some excellent findings have been explored for the development of MOF as a heterogeneous catalyst, enough space is left for further improvement/discoveries, such as use of the same framework for formation of both enantiomers selectively depending upon suitable stimuli such as heat, light, anion or solvent polarity, etc. under ambient conditions.

REFERENCES

1. Wright, P. A. (2008). *Microporous Framework Solids*, (pp. 1–7). Royal Society of Chemistry, Cambridge.
2. (a) Kondo, M.; Yoshitomi, T.; Matsuzaka, H.; Kitagawa, S.; Seki, K. (1997). Kondo, M.; Yoshitomi, T.; Matsuzaka, H.; Kitagawa, S.; Seki, K. (1997). Three-Dimensional Framework with Channeling Cavities for Small Molecules: {[M2(4,4′-bpy)$_3$(NO$_3$)$_4$]·xH$_2$O}$_n$(M = Co, Ni, Zn). *Angew. Chem., Int. Ed. Engl., 36*(16), 1725–1727. (b) Cheetham, A. K.; Férey, G.; Loiseau, T. (1999). Open framework Inorganic materials. *Angew. Chem., Int. Ed., 38*(22), 3268–3292. (c) Furukawa, H.; Cordova, K. E.; O'Keeffe, M.; Yaghi, O. M. (2013). The chemistry and applications of metal-organic framework. *Science, 341*, 974–986.

3. Côté, A. P.; Benin, A. I.; Ockwig, N. W.; O'Keeffe, M.; Matzger, A. J.; Yaghi O. M. (2005). Porous, crystalline, covalent organic frameworks. *Science*, *310*, 1166–1170.
4. Thomas, A. (2010). Functional materials: From hard to soft porous frameworks. *Angew. Chem., Int. Ed.*, *49*, 8328–8344.
5. Li, H.; Eddaoudi, M.; O'Keeffe, M.; Yaghi, O. M. (1999). Design and synthesis of an exceptionally stable and highly porous metal-organic framework. *Nature*, *402*, 276–279.
6. Lan, Y. Q.; Jiang, H. L.; Li, S. L.; Xu, Q. (2011). Mesoporous metal-organic frameworks with size-tunable cages: selective CO_2 uptake, encapsulation, of Ln^{3+} cations for luminiscene and column chromatographic dye separation. *Adv. Mater.*, *23*, 5015–5020.
7. Dybtsev, D. N.; Chun, H.; Kim, K. (2004). Rigid and flexible: a highly porous metal-organic framework with unusual guest-dependent dynamic behavior. *Angew. Chem., Int. Ed.*, *43(38)*, 5033–5036.
8. Matsuda, R.; Kitaura, R.; Kitagawa, S.; Kubota, Y.; Belosludov, R. V.; Kobayashi, T. C.; Sakamoto, H.; Chiba, T.; Takata, M.; Kawazoe, Y.; Mita, Y. (2005). Highly controlled acetylene accommodation in metal-organic micro porous material. *Nature*, *436*, 238–241.
9. Kubota, Y.; Takata, M.; Matsuda, R.; Kitaura, R.; Kitagawa, S.; Kato, K.; Sakata, M.; Kobayashi, T. C. (2005). Cover picture: direct observation of hydrogen molecules adsorbed onto a microporous coordination polymer. *Angew. Chem., Int. Ed.*, *44(6)*, 829.
10. Ma, S.; Zhou, H.-C. (2006). A metal-organic framework with entatic metal centres exhibiting high gas adsorption affinity. *J. Am. Chem. Soc.*, *128*, 11734–11735.
11. Murray, L. J.; Dinca, M.; Long, J. R. (2009). Hydrogen storage in metal organic-frameworks. *Chem. Soc. Rev.*, *38*, 1294–1314.
12. Furukawa, H.; Ko, N.; Go, Y. B.; Aratani, N.; Choi, S. B.; Choi, E.; Yazaydin, A. Ö.; Snurr, R. Q.; O'Keeffe, M.; Kim, J.; Yaghi, O. M. (2010). Ultrahigh porosity in metal-organic frameworks. *Science*, *329*, 424–428.
13. Pan, L.; Adams, K. M.; Hernandez, H. E.; Wang, X.; Zheng, C.; Hattori, Y.; Kaneko, K. (2003). Porous lanthanide – organic frameworks: synthesis, characterization and unprecedented gas adsorption properties. *J. Am. Chem. Soc.*, *125*, 3062.
14. Dybtsev, D. N.; Chun, H.; Yoon, S. H.; Kim, D.; Kim, K. (2004). Microporous manganese formate: a simple metal-organic porous material with high framework stability and highly selective gas sorption properties. *J. Am. Chem. Soc.*, *126*, 32–33.
15. Hayashi, H.; Cote, A. P.; Furukawa, H.; O'Keeffe, M.; Yaghi, O. M. (2007). Zeolite a imidazolate frameworks. *Nat. Mater.*, *6*, 501–506.
16. Samsonenko, D. G.; Kim, H.; Sun, Y.; Kim, G.-H.; Lee, H.-S.; Kim, K. (2007). Microporous magnesium and manganese formates for acetylene storage and separation. *Chem. Asian. J.*, *2*, 484–488.
17. Kim, H.; Samsonenko, D. G.; Yoon, M.; Yoon, J. W.; Hwang, Y. K.; Chang, J.-S.; Kim, K. (2008). Temperature-triggered gate opening for gas adsorption in microprous manganese fomrtae. *Chem. Commun,*, *(39)*, 4697–4699.
18. Li, J.-R.; Kuppler, R. J.; Zhou, H.-C. (2009). Selective gas adsorption and separation in metal-organic frameworks. *Chem. Soc. Rev.*, *38*, 1477–1504.
19. Chen, B.; Xiang, S.; Qian, G. (2010). Metal-organic frameworks with functional pores for recognition of small molecules. *Acc. Chem. Res.*, *43(8)*, 1115–1124.
20. Jiang, H.-L.; Xu, Q. (2011). Porous metal-organic frameworks as platform for functional applications. *Chem. Commun.*, *47*, 3351–3370.
21. Li, J.-R.; Sculley, J.; Zhou, H.-C. (2011). Metal–organic frameworks for separations. *Chem. Rev.*, *112(2)*, 869–932.
22. Qiu, Y. C.; Deng, H.; Mou, J. X.; Zeller, M.; Batten, S. R.; Wu, H. H.; Li, J. (2009) *Chem. Commun.*, 5415.

23. Allendorf, M. D.; Bauer, C. A.; Bhakta, R. K.; Houk, R. J. T. (2009). Luminiscent metal-organic frameworks. *Chem. Soc. Rev.*, *38*, 1330–1352.

24. Achmann, S.; Hagen, G.; Kita, J.; Malkowsky, I. M.; Kiener, C.; Moos, R. (2009). Meta–organic frameworks for sensing applications in the gas phase. *Sensor*, *9*, 1574.

25. Jiang, H.-L.; Tatsu, Y.; Lu, Z.-L.; Xu, Q. (2010). Non-, micro-, and mesoporous metal-organic framework isomers: reversible transformation, fluorescence sensing and large molecule separation. *J. Am. Chem. Soc.*, *132(16)*, 5586–5587.

26. Cui, Y.; Yue, Y.; Qian, G.; Chen, B. (2011). Luminescent functional metal–organic frameworks. *Chem. Rev.*, *112(2)*, 1126–1162.

27. Kahn, O. (1993). *Molecular Magnetism*, VCH: Weinheim.

28. Halder, G. J.; Kepert, C. J.; Moubaraki, B.; Murray, K. S.; Cashion, J. D. (2002). *Science*, *298*, 1762.

29. Kepert, C. J. (2006). Advanced functional properties in nanoporous coordination frameworks materials. *Chem. Commun.*, *7*, 695–700.

30. Rao, C. N. R.; Cheetham, A. K.; Thirumurugan, A. (2008). Hybrid inorganic-organic materials: a new family in condensed matter physics. *J. Phys.: Condens. Matter*, *20*, 083202.

31. Kurmoo, M. (2009). Magnetic metal–organic framework. *Chem. Soc. Rev.*, *38*, 1353–1379.

32. Kim, H.; Sun, Y.; Kim, Y.; Kajiwara, T.; Yamashita, M.; Kim, K. (2011). Metal-organic frameworks with rare topologies:ionsdaleite – type metal formates and their magnetic properties. *CrystEngComm*, *13*, 2197–2200.

33. Zhang, W.; Xiong, R.-G. (2011). Ferroelectric metal–organic frameworks. *Chem. Rev.*, *112(2)*, 1163–1195.

34. Fujita, M.; Kwon, Y. J.; Washizu, S.; Ogura, K. (1994). Preparation, clathration ability and catalysis of two-dimensional square network material composed of cadmium(II) and 4,4′-bipyridine. *J. Am. Chem. Soc.*, *116*, 1151–1152.

35. Hasegawa, S.; Horike, S.; Matsuda, R.; Furukawa, S.; Mochizuki, K.; Kinoshita, Y.; Kitagawa, S. (2007). Three dimensional porous coordination polymer functionalized with amide groups based on tridentate ligand: selective sorption and catalysis. *J. Am. Chem. Soc.*, *129*, 2607–2614.

36. Alkordi, M. H.; Liu, Y. L.; Larsen, R. W.; Eubank, J. F.; Eddaoudi, M. (2008). Zeolite metal-like organic frameworks as platform for organic applications: on metallophorphy-rin based catalysts. *J. Am. Chem. Soc.*, *130*, 12639–12641.

37. Horike, S.; Dinca, M.; Tamaki, K.; Long, J. R. (2008). Size selective lewis acid catalysis in microporous metal-organic framework with exposed Mn^{2+} coordination sites. *J. Am. Chem. Soc.*, *130*, 5854.

38. Lee, J.-Y.; Farha, O. M.; Roberts, J.; Scheidt, K. A.; Nguyen, S. T.; Hupp, J. T. (2009). Metal-organic framework materials as catalyst. *Chem. Soc. Rev.*, *38*, 1450–1459.

39. Farrusseng, D.; Aguado, S.; Pinel, C. (2009). Metal-organic frameworks: opportunities for catalysis. *Angew. Chem., Int. Ed.*, *48*, 7502–7513.

40. Corma, A.; Garcia, H.; Llabres, F. X.; Xamena, I. (2010). Engineering metal-organic frameworks for heterogeneous catalysis. *Chem. Rev.*, *110*, 4606–4655.

41. Jiang, H.-L.; Akita, T.; Ishida, T.; Haruta, M.; Xu, Q. (2011). Synergistic catalysis of Au@Ag core-shell nanoparticles stabilized on metal-organic frameworks. *J. Am. Chem. Soc.*, *133*, 1304–1306.

42. (a) Fiedler, D.; Bergman, R. G.; Raymond, K. N. (2006). Stabilization of reactive organometallic intermediates inside a self-assembled nanoscale host. *Angew. Chem., Int. Ed.*, *45*, 745–748. (b) Pluth, M. D.; Bergman, R. G.; Raymond, K. N. (2007). Catalytic deprotection of acetals in basic solutions with self-assembled supramolecular

"nanozyme". *Angew. Chem., Int. Ed.*, *46*, 8587–8589. (c) Parac, T. N.; Caulder, D. L.; Raymond, K. N. (1998). Selective encapsulation of aqueous cationic guests into a supramolecular tetrahedral $[M_4L_6]^{12-}$ anionic guest. *J. Am. Chem. Soc.*, *120*, 8003–8004. (d) Pluth, M. D.; Bergman, R. G.; Raymond, K. N. (2007). Making amines strong bases: thermodynamic stabilization of protonated guests in highly – charged supramolecular host. *J. Am. Chem. Soc.*, *129*, 11459–11467.

43. Yoshizawa, M.; Tamura, M.; Fujita M. (2006). Diels-alder in aqueous molecular host: unusual regioselectivity and efficient catalysis. *Science*, *312*, 251–254.

44. Eddaoudi, M.; Li, H.; Reineke, T.; Fehr, M.; Kelley, D.; Groy, T. L.; Yaghi, O. M. (1999). Design and synthesis of metal-carboxylate frameworks with permanent microporosity. *Top. Catal.*, *9 (1–2)*, 105–111.

45. Nugent, W. A.; RajanBabu, T. V.; Burk, M. (1999). Beyond nature's chiral pool: enantioselective catalysis in industry. *J. Science*, *259*, 479–483.

46. Jones, C. W.; McKittrick, M. W.; Nguyen, J. V.; Yu, K. (2005). Design of silica tethered metal complexes for polymerization catalysis. *Top. Catal.*, *34*, 67–76.

47. Mizuno, N.; Misono, M. (1998). Heterogeneous catalysis. *Chem. Rev.*, *98*, 199–218.

48. Hattori, H. (1995). Heterogeneous basic catalysis. *Chem. Rev.*, *95*, 537.

49. Yaghi, O. M.; O'Keeffe, M.; Ockwig, N. W.; Chae, H. K.; Eddaoudi, M.; Kim, J. (2003). Reticular synthesis and the design of new materials. *Nature*, *423*, 705–714.

50. Tranchemontagne, D. J.; Mendoza-Cortés, J. L.; O'Keeffe, M.; Yaghi, O. M. (2009). Secondary building units, nets and bonding chemistry of metal-organic frameworks. *Chem. Soc. Rev.*, *38*, 1257.

51. Eddaoudi, M.; Moler, D. B.; Li, H.; Chen, B.; Reineke, T. M.; O'Keeffe, M.; Yaghi, O. M. (2001). Modular chemistry: secondary building units as basis for design of highly porous and robust metal-organic carboxylate frameworks. *Acc. Chem. Res.*, *34*, 319–330.

52. Moulton, B.; Zaworotko, M. J. (2001). From molecules to crystal engineering: supramolecular isomerism and polymorphism in network solids. *Chem. Rev.*, *101*, 1629.

53. Zaworotko, M. J. (2001). Superstructural diversity in two dimension: crystal engineering of laminated solids. *Chem. Commun.*, *1*, 1–9.

54. O'Keeffe, M.; Yaghi, O. M. (2011). Deconstructing the crystal structures of metal–organic frameworks and related materials into their underlying nets. *Chem. Rev.*, *112(2)*, 675–702.

55. Xu, W.; Thapa, K. B.; Ju, Q.; Fang, Z.; Huang, W. (2018). *Coord. Chem. Rev.*, *373*, 199–232.

56. Yoon, M.; Srirambalaji, R.; Kim, K. (2012). Homochiral metal-organic framework for asymmetric heterogeneous catalysis. *Chem. Rev.*, *112*, 1196–1231.

57. Nelson, A. P.; Farha, O. K.; Mulfort, K. L.; Hupp, J. T. (2009). Supercritical processing as a route to high internal surface areas and permanent microsporosity in metal-organic framework materials. *J. Am. Chem. Soc.*, *131*, 458.

58. Ma, L.; Jin, A.; Xie, Z.; Lin, W. (2009). Freeze dying significantly increase permanent porosity and hydrogen uptake in 4,4′-connected metal-organic frameoworks. *Angew. Chem., Int. Ed.*, *48*, 9905–9908.

59. Férey, G.; Mellot-Draznieks, C.; Serre, C.; Millange, F.; Dutour, J.; Surblé, S.; Margiolaki, I. (2005). A chromic terephthalate – based solid with unusually large pore volumes and surface areas. *Science*, *309*, 2040–2042.

60. Banerjee, M.; Das, S.; Yoon, M.; Choi, H. J.; Hyun, M. H.; Park, S. M.; Seo, G.; Kim, K. (2009). Postsynthetic modification switches an achiral framework to catalytically active homochiral metal-organic porous materials. *J. Am. Chem. Soc.*, *131*, 7524–7525.

61. Fujita, M.; Kwon, Y. J.; Washizu, S.; Ogura, K. (1994). Preparation, clatharation ability, and catalysis of two-dimensional square network material composed of cadmium(II) and 4,4'-bipyridine. *J. Am. Chem. Soc.*, *116 (3)*, 1151–1152.
62. Seo, J. S.; Whang, D.; Lee, H.; Jun, S. I.; Oh, J.; Jeon, Y.; Kim, K. (2000). *Nature*, *404*, 982.
63. Evans, O. R.; Ngo, H. L.; Lin, W. (2001). Chiral porous solids based on lamellar lanthanide phosphate. *J. Am. Chem. Soc.*, *123(42)*, 10395–10396.
64. Dhakshinamoorthy, A.; Alvaro, M.; Garcia, H. (2011). Metal-organic frameworks as heterogeneous catalysts for oxidation reactions. *Catal. Sci. Technol.* **2**, *1*, 856–867.
65. Dybtsev, D. N.; Nuzhdin, A. L.; Chun, H.; Bryliakov, K. P.; Talsi, E. P.; Fedin, V. P.; Kim, K. (2006). A homochiral metal-organic material with permanent porosity, enantioselective sorption properties and catalytic activities. *Angew. Chem., Int. Ed.*, *45*, 916–920.
66. Hwang, Y. K.; Hong, D.-Y.; Chang, J.-S.; Seo, H.; Yoon, M.; Kim, J.; Jhung, S. H.; Serre, C.; Férey, G. (2009). Selective sulfoxidation of aryl halides by coordinatively unsaturated metal-centres in chromium carboxylate MIL-101. *Appl. Catal. A*, *358(2)*, 249–253.
67. Dybtsev, D. N.; Yutkin, M. P.; Samsonenko, D. G.; Fedin, V. P.; Nuzhdin, A. L.; Bezrukov, A. A.; Bryliakov, K. P.; Talsi, E. P.; Belosludov, R. V.; Mizuseki, H.; Kawazoe, Y.; Subbotin, O. S.; Belosludov, V. R. (2010). Modular, homochiral, porous coordination polymers: rational design, enantioselective guest exchange sorption and ab initio calculations of host –guest interactions. *Chem. - Eur. J.*, *34*, 10348.
68. Brown, K.; Zolezzi, S.; Aguirre, P.; Venegas-Yazigi, D.; Paredes-Garcia, V.; Baggio, R.; Novak, M. A.; Spodine, E. (2009). [Cu(H$_2$btec)(bipy)]$^\infty$: a novel metal organic framework (MOF) as heterogeneous catalyst for the oxidation of olefins. *Dalton Trans.*, *(8)*, 1422–1427.
69. Sheldon, R. A. (1980) Synthetic and mechanistic aspects of metal-catalysed epoxidations with hydroperoxides *J. Mol. Catal.* 7, 107.
70. Carrell, T. G.; Cohen, S.; Dismukes, G. (2002). Oxidative catalysis by Mn$_4$O$_4$$^{6+}$ cubane complexes. *J. Mol. Catal. A*, *187*, 3–15.
71. Mukherjee, S.; Samanta, S.; Bhaumik, A.; Ray, B. C. (2006). Mechanistic study of cyclohexene oxidation and its use in modification of industrial waste organics. *Appl. Catal. B*, *68*, 12.
72. Schlichte, K.; Kratzke, T.; Kaskel, S. (2004). Improved synthesis, thermal stability and catalytic properties of metal-organic framework compound Cu$_3$(BTC)$_2$. *Micropor. Mesopor. Mater.*, *73*, 81–88.
73. Marx, S.; Kleist, W.; Baiker, A. (2011). Synthesis, structural properties and acatlytic behavior of Cu-BTC and mixed linker Cu-BTC-PyDC in the oxidation of benzene derivatives. *J. Catal.*, *281*, 76–87.
74. Ingleson, M.; Barrio, J.; Bacsa, J.; Dickinson, C.; Park, H.; Rosseinsky, M. (2008). Generation of solid Brønsted acid site in chiral framework. *Chem. Commun.*, *(11)*, 1287–1289.
75. Su, Y.; Chang, G.; Zhang, Z.; Xing, H.; Su, B.; Yang, Q.; Ren, Q.; Yang, Y.; Bao, Z. (2016). Catalytic dehydration of glucose to 5-hydroxymethylfurfural with a bifunctional metal-organic framework. *AIChE J.*, *62*, 4403–4417.
76. Qi, X; Guo, H.; Li, L.; Smith, R. L. (2012). Acid catalyzed dehydration of fructose into 5-hydroxylmethylfurfural by cellulose derived-amorphous carbon. *ChemSusChem.*, *5*, 2215–2220.
77. Xie, H.; Zhao, Z. K.; Wang, Q. (2012). Catalytic conversion of inulin and fructose into 5-hydroxylmethylfurfural by lignosulphonic acid in ionic liquids. *ChemSusChem.*, *5*, 901–905.

78. Kotadia, D. A.; Soni, S. S. (2013). Symmetrical and unsymmetrical Brønsted acidic ionic liquids for the effective conversion of fructose to 5-hydroxymethylfurfural. *Catal Sci Technol.*, *3*, 469–474.

79. van Putten, R. J.; van der Waal, J. C.; de Jong, E.; Rasrendra, C. B.; Heeres, H. J.; de Vries, J. G. (2013). Hydroxymethylfurfural, a versatile platform chemical made from renewable resources. *Chem. Rev.*, *113*, 1499–1597.

80. Ono, Y.; Baba, T. (1997). Selective reactions over solid base catalysts. *Catal. Today*, *38*, 321–337.

81. Moreau, C.; Durand, R.; Roux, A.; Tichit, D. (2000). Isomerization of glucose into fructose in presence of cation-exchanged zeolites and hydrotalcites. *Appl. Catal. A Gen.*, *193*, 257–264.

82. Moliner, M.; Roman-Leshkov, Y.; Davis, M. E. (2010). Tin-containing zeolites are highly active catalysts for the isomerization of glucose in water. *Proc. Natl. Acad. Sci. USA*, *107*, 6164–6168.

83. Kesanli, B.; Cui, Y.; Smith, M. R.; Bittner, E. W.; Bockrath, B. C.; Lin, W. (2005). Highly interpenetrated metal-organic frameworks for hydrogen storage. *Angew. Chem., Int. Ed.*, *44*, 72–75.

84. (a) Yang, J. W.; Chandler, C.; Stadler, M.; Kampen, D.; List, B. (2008). Proline catalyzed-Mannich reactions of acetaldehyde. *Nature*, *452*, 453–455. (b) List, B.; Lerner, R. A.; Barbas III, C. F. (2000). Proline catalyzed direct asymmetric aldol reactions. *J. Am. Chem. Soc.*, *122*, 2395. (c) Wolfgang, N.; Fujie, T.; Barbas III, C. F. (2004). Enamine based organocatalysis with prolines and diamines: the development of direct catalytic asymmetric aldol, Mannich, Michael and Diels alder reaction. *Acc. Chem. Res.*, *37*, 580–591. (d) D'Elia, V.; Zwicknagl, H.; Reiser, O. (2008). Short α/β peptides as catalysts for intra and inter molecular aldol reactions. *J. Org. Chem.*, *73*, 3262–3265.

85. Gedrich, K.; Heitbaum, M.; Notzon, A.; Senkovska, I.; Frçhlich, R.; Getzschmann, J.; Mueller, U.; Glorius, F.; Kaskel, S. (2011). *Chem. - Eur. J.*, *17*, 2099–2106.

86. Lin, X.; Jia, J.; Zhao, X.; Thomas, K. M.; Blake, A. J.; Walker, G. S.; Champness, N. R.; Hubberstey, P.; Schroeder, M. (2006). High H_2 adsorption by coordination framework materials. *Angew. Chem.*, *118*, 7518–7524; *Angew. Chem. Int. Ed.*, **2006**, *45*, 7358–7364.

87. Lin, X.; Telepeni, I.; Blake, A. J.; Dailly, A.; Brown, C. M.; Simmons, J. M.; Zoppi, M.; Walker, G. S.; Thomas, K. M.; Mays, T. J.; Hubberstey, P.; Champness, N. R.; Schroder, M. (2009). High capacity hydrogen adsorption in Cu(II) tetracarboxylate framework materials: the role of pore size, ligand functionalization and exposed metal sites. *J. Am. Chem. Soc.*, *131*, 2159–2171.

88. Schlichte, K.; Kratzke, T.; Kaskel, S. (2004). Improved synthesis, thermal stability catalytic properties of metal-organic framework compound $Cu_3(BTC)_2$. *Microporous Mesoporous Mater.*, *73*, 81–88.

89. Chae, H. K.; Siberio-Perez, D. Y.; Kim, J.; Go, Y. B.; Eddaoudi, M.; Matzger, A. J.; O'Keeffe, M.; Yaghi, O. M. (2004). A route to high surface area, porosity and inclusion of large molecules in crystals. *Nature*, *427*, 523–527.

90. Tanabe, K. K.; Cohen, S. M. (2009). Engineering a metal-organic framework catalyst by using postsynthetic modification. *Angew. Chem., Int. Ed.*, *48*, 7424–7427.

91. Kitagawa, S.; Noro, S.-I.; Nakamura, T. (2006) Pore surface engineering of microporous coordination. *Chem. Commun.*, 701–707.

92. De Wu, C.; Hu, A.; Zhang, L.; Lin, W. (2005). A homochiral porous metal-organic framework for highly enantioselective heterogeneous asymmetric catalysis. *J. Am. Chem. Soc.*, *127*, 8940–8941.

93. Wu, C. D.; Lin, W. (2007). Heterogeneous asymmetric catalysis with homochiral metal-organic frameworks: network-structure-dependent catalytic activity. *Angew Chem., Int. Ed.*, *46*, 1075–1078.

94. Dalton, C. T.; Ryan, K. M.; Wall, V. M.; Bousquet, C.; Giheany, D. G. (1998) Recent progress towards the understanding of metal–salen catalysed asymmetric alkene epoxidation. *Top Catal. 5*, 75–91.

95. Cho, S.-H.; Ma, B.; Nguyen, S. T.; Hupp, J. T.; Albrecht-Schmitt, T. E. (2006). A metal-organic framework material that functions as an enantioselective catalyst for olefinic epoxidation. *Chem. Commun.*, 2563–2565.

96. (a) Katsuki, T. (1995). Catalytic asymmetric oxidations using optically active (salen) manganese (III) complexes as catalysts. *Coord. Chem. Rev.*, *140*, 189–214. (b) Palucki, M.; Finney, N. S.; Pospisil, P. J.; Guler, M. L.; Ishida T.; Jacobsen, E. N. (1998). The mechanistic basis for electronic effects on enantioselectivity in the (salen) Mn (III) – catalyzed epoxidation reaction. *J. Am. Chem. Soc.*, *120*, 948–954.

97. Shultz, A. M.; Farha, O. K.; Hupp, J. T.; Nguyen, S. T. (2009). A catalytically active, permanently microporous MOF with metalloporphyrin struts. *J. Am. Chem. Soc.*, *131*, 4204–4205.

98. Ma, B. Q.; Mulfort, K. L.; Hupp, J. T. (2005). Microporous pillared paddle-wheel frameworks based on metal-ligand coordination of zinc ions. *Inorg. Chem.*, *44*, 4912–4914.

99. Mackay, L. G.; Wylie, R. S.; Sanders, J. K. M. (1994). Catalytic acyl transfer by a cyclic porphyrin trimer: efficient turnover without product inhibition. *J. Am. Chem. Soc.*, *116*, 3141–3142.

100. Oliveri, C. G.; Gianneschi, N. C.; Nguyen, S. T.; Mirkin, C. A.; Stern, C. L.; Wawrzak, Z.; Pink, M. (2006). Supramolecular allosteric cofacial porphyrin complexes. *J. Am. Chem. Soc.*, *128*, 16286–16296.

101. (a) Sheldon, R. A.; Kochi, J. K. (1981). *Metal-Catalyzed Oxidations of Organic Compounds*, Academic Press: New York. (b) *Metalloporphyrins in Catalytic Oxidations* (1994). R. A. Sheldon (Ed.), Marcel Dekker: New York. (c) Collman, J. P.; Zhang, X.; Lee, V. J.; Uffelman, E. S.; Brauman, J. I. (1993). Regioselective and enantiose-lective epoxidation catalyzed by metalloporphyrins. *Science*, *261*, 1404–1411. (d) *Comprehensive Supramolecular Chemistry* (1996). K. S. Suslick (Ed.), Supramolecular Reactivity and Transport: Bioinorganic Systems, Pergamon: Oxford. (e) Meunier, B. (1994). In *Metalloporphyrin Catalyzed Oxidations*. F. Montanari; L. Casella (Eds.), (pp. 1–48). Kluwer: Dordrecht, The Netherlands. (f) Shultz, Abraham M; Farha, Omar K; Hupp, Joseph T; Nguyen, SonBinh T (2009) A catalytically active, permanently micro-porous MOF with metalloporphyrin struts, *J. Am. Chem. Soc.*, 131 (12), 4204–4205.

102. Guo, C.; Song, J.; Chen, X.; Jiang, G. (2000). A new evidence of the high valent-oxo-metal radical cation intermediate and hydrogen radical abstract mechanism in hydrocarbon-hydroxylation catalyzed by metalloporphsrins. *J. Mol. Catal. A.*, *157*, 31–40.

103. Sacco, H. C.; Iamamoto, Y.; Lindsay Smith, J. R. (2001). Alkeneepoxidation with iodo-sylbenzene catalysed by polyionic manganese porphyrins electrostatically bound to counter-charged supports *J. Chem. Soc. Perkin Trans. 2*, 181–190.

104. Vinhado, F. S.; Prado-Manso, C. M. C.; Sacco, H. C.; Iamamoto, Y. (2001). Cationic manganese(III) porphyrins bound to novel bis-functionalised silica as catalysts for hydrocarbons oxygenation by iodosylbenzene and hydrogen peroxide. *J. Mol. Catal. A*, *174*, 279–288.

105. Deniaud, D.; Spyroullias, G. A.; Bartoli, J. F.; Battioni, P.; Mansuy, D.; Pinel, C.; Odobel, F.; Bujoli, B. (1998). Shape selectivity for alkane hydroxylation with a new class of phosphonate – based heterogenised manganese porphyrins. *New. J. Chem.*, *22*, 901–905.

106. Li, Z.; Xia, C. G.; Zhang, X. M. (2002). Preparation and catalysis of DMY and MCM-41 encapsulated cationic Mn(III)–porphyrin complex. *J. Mol. Catal. A.*, *185*, 47–56.

107. (a) Khan, T. A.; Hriljac, J. A. (1999). Hydrothermal synthesis of microporous materials with direct incorporation of porphyrin molecules. *Inorg. Chim. Acta.*, *294*, 179–182. (b) Rosa, I. L. V.; Manso, C. M. C. P.; Serra, O. A.; Iamamoto, Y. (2000). Biomemetical catalytic activity of iron(III) porphyrins encapsulated in zeolite X. *J. Mol. Catal. A*, *160*, 199–208. (c) Skrobot, F. C.; Rosa, I. L. V.; Marques, A. P. A.; Martins, P. R.; Rocha, J.; Valente, A. A.; Iamamoto, Y. (2005). Asymmetric cationic methyl pyridyl and pentafuorophenyl porphyrin encapsulated in zeolites: a cytochrome P-450 model. *J. Mol. Catal. A*, *237*, 86–92.

108. (a) Xu, W.; Guo, H.; Akins, D. L. (2001). Aggregation of Tetrakis (p-sulfonatophenyl) porphyrin within modified mesoporous MCM-41. *J. Phys. Chem. B*, *105*, 1543–1546. (b) Nur, H.; Hamid, H.; Endud, S.; Hamdan, H.; Ramli, Z. (2006). Iron-porphyrin encapsulated in poly(methacrylic acid) and mesoporous Al-MCM-41 as catalysts in the oxidation of benzene to phenol. *Mater. Chem. Phys.*, *96*, 337–342.

109. Moreira, M. S. M.; Martins, P. R.; Curi, R. B.; Nascimento, O. R.; Iamamoto, Y. (2005). Iron porphyrins immobilized on silica surface and encapsulated in silica matrix: a comparison of their catalytic activity in hydrocarbon oxidation. *J. Mol. Catal. A*, *233*, 73–81.

110. Canioni, R.; Roch-Marchal, C.; Sécheresse, F.; Horcajada, P.; Serre, C.; Hardi-Dan, M.; Férey, G.; Grenèche, J.-M.; Lefebvre, F.; Chang, J.-S.; Hwang, Y.-K.; Lebedev, O.; Turner, S.; Van Tendeloo, G. (2011). *J. Mater. Chem.*, *21*, 1226–1233.

111. Zhang, F.; Jin, Y.; Shi, J.; Zhong, Y.; Zhu, W.; El-Shall, M. S. (2015). Polyoxometalates confined in the mesoporous cages of metal–organic framework MIL-100(Fe): efficient heterogeneous catalysts for esterification and acetalization reactions. *Chem. Eng. J. 269*, 236–244.

112. Zhu, Z.; Tain, R.; Rhodes, C. (2003). A study of decomposition behaviour of 12-tungstophosphate heteropolyacid in solution. *Can. J. Chem.*, *81*, 1044–1050.

113. Chen, M.; Yan, J.; Tan, Y.; Li, Y.; Wu, Z.; Pan, L.; Liu, Y. (2015). Hydroxyalkylation of phenol with formaldehyde to bisphenol F catalyzed by keggin phosphotungstic acid encapsulated in metal-organic frameworks MIL-100(Fe or Cr) and MIL-101 (Fe or Cr). *Ind. Eng. Chem. Res.*, *54*, 11804–11813.

114. Hermes, S.; Schroter, M. K.; Schmid, R.; Khodeir, L.; Muhler, M.; Tissler, A.; Fischer, R. W.; Fischer, R. A. (2005). Metal@ MOF: loading of highly porous coordination polymers host lattices by metal organic chemical vapor deposition. *Angew. Chem., Int. Ed.*, *44*, 6237–6241.

115. Müller, M.; Hermes, S.; Kähler, K.; van den Berg, M. W. E.; Muhler, M.; Fischer, R. A. (2008). Loading of MOF-5 with Cu and ZnO nanoparticles by gas-phase infiltration with organometallic precursors: properties of Cu/ZnO @ MOF-5 as catalyst for methanol synthesis. *Chem. Mater.*, *20*, 4576–4587.

116. Schröder, F.; Esken, D.; Cokoja, M.; van den Berg, M. W.; Lebedev, O. I.; Van Tendeloo, G.; Walaszek, B.; Buntkowsky, G.; Limbach, H. H.; Chaudret, B.; Fischer, R. A. (2008). Ruthenium nanoparticles inside porous $[Zn_4O(bdc)_4]$ by hydrogenolysis of adsorbed [Ru(cod)(cot)]: a solid state reference system for surfactant stabilized Ruthenium colloids. *J. Am. Chem. Soc.*, *130*, 6119–6130.

117. Wu, X. Q.; Huang, D. D.; Zhou, Z. H.; Dong, W. W.; Wu, Y. P.; Zhao, J.; Li, D. S.; Zhang, Q. C.; Bu, X. H. (2017). Ag-NPs embedded in two novel Zn_3/Zn_5 cluster based metal-organic frameworks for catalytic reduction of 2/3/4-nitrophenol. *Dalton Trans.*, *46*, 2430–2438.

118. Qin, L.; Li, Z. W.; Xu, Z. H.; Guo, X. W.; Zhang, G. L. (2015). Organic-acid-directed assembly of iron-carbon oxide nanoparticles on coordinatively unsaturated metal sites of MIL-101 for green photochemical oxidation. *Appl. Catal. B-Environ.*, *179*, 500–508.

119. Zhang, Y. P.; Zhou, Y.; Zhao, Y.; Liu, C. J. (2016). Recent progresses in the size and structure control of MOF supported noble metal catalysts. *Catal. Today*, *263*, 61–68.

120. Yu, J.; Mu, C.; Yan, B. Y.; Qin, X. Y.; Shen, C.; Xue, H. G.; Pang, H. (2017). Nanoparticle/ MOF composites: preparations and applications. *Mater. Horiz.*, *4*, 557–569.

121. Jiang, H. L.; Liu, B.; Akita, T.; Haruta, M.; Sakurai, H.; Xu, Q. (2009). Au@ ZIF-8: CO oxidation over gold nanoparticles deposited to metal-organic frameworks. *J. Am. Chem. Soc.*, *131*, 11302–11303.

122. Dhakshinamoorthy, A.; Garcia, H. (2012). Catalysis by metal nanoparticles embedded on metal-organic frameworks. *Chem. Soc. Rev.*, *41*, 5262–5284.

123. Zhao, M.; Yuan, K.; Wang, Y.; Li, G.; Guo, J.; Gu, L.; Hu, W.; Zhao, H.; Tang, Z. (2016). Metal-organic frameworks as selectively regulators for hydrogenation reactions. *Nature*, *539*, 76–80.

124. Ide, M. S.; Hao, B.; Neurock, M.; Davis, R. J. (2012). Mechanistic insights on the hydrogenation of α,β-unsaturated ketones and aldehydes to unsaturated alcohols over metal catalysts. *ACS Catal.*, *2*, 671–683.

125. Tian, Z.; Xiang, X.; Xie, L.; Li, F. (2013). Liquid-phase hydrogenation of cinnamalde-hdye: Enhancing selectivity of supported gold catalysts by incorporation of Cerium into the support. *Ind. Eng. Chem. Res.*, *52*, 288–296.

126. Kahsar, K. R.; Schwartz, D. K.; Medlin, J. W. (2014). Control of metal catalyst selec-tivity through specific non covalent molecular interactions. *J. Am. Chem. Soc.*, *136*, 520–526.

127. Wu, B.; Huang, H.; Yang, J.; Zheng, N.; Fu, G. (2012). Selective hydrogenation of α,β-unsaturated aldehydes catalyzed by amine – capped platinum-cobalt nanocrystals. *Angew. Chem., Int. Ed.*, *51*, 3440–3443.

128. Stassi, J. P.; Zgolicz, P. D.; de Miguel, S. R.; Scelza, O. A. (2013). Formation of different promoted metallic phases in PtFe and PtSn catalysts supported on carbonaceous materials used for selective hydrogenation. *J. Catal.*, *306*, 11–29.

129. Kennedy, G.; Baker, L. R.; Somorjai, G. A. (2014). Selective amplification of C=O bond hydrogenation on Pt/TiO$_2$: catalytical reaction and sum-frequency generation vibra-tional spectroscopy studies of crotonaldehyde hydrogenation. *Angew. Chem., Int. Ed.*, *53*, 3405–3408.

130. Kliewer, C. J.; Bieri, M.; Somorjai, G. A. (2009). Hydrogenation of the α,β-unsaturated aldehydes acrolein, crotonaldehyde, and prenal over Pt single crystals: a kinetic and Sum-frequency generation vibrational spectroscopy studies. *J. Am. Chem. Soc.*, *131*, 9958–9966.

131. Canivet, J.; Vandichel, M.; Farrusseng, D. (2016). Origin of highly active metal-organic framework catalysts: defects? Defects! *Dalton Trans.*, *45*, 4090–4099.

7 Role of Metal-heterogeneous Catalysts in Organic Synthesis

Sukhdev Singh
iSm2 *Stereo*, Aix-Marseille University, Marseille, France

Deepika
Chaudhary Devi Lal University, Sirsa (Haryana), India

CONTENTS

7.1 HETEROGENEOUS CATALYSTS IN ORGANIC SYNTHESIS

The use of homogeneous catalysts in organic synthesis is well established. However, there are a few disadvantages associated with the usage of homogeneous catalysts in synthetic chemistry. The prominent ones are the formation of soluble complex, especially at a large scale, the problem of separation from the reaction mixture and lack of reusability and toxicity issues that pose health risks. On the other hand, heterogeneous catalysts provide a great advantage over homogeneous catalysts. They are easy to use, cheap and reusable and, their handling and separation from the reaction mixture are very simple. Moreover, they are less toxic, and more stable than homogeneous catalysts. Heterogeneous catalysts come in different forms like pure metals (especially transition metals, metal powders, etc.), metals combined with some other components such as metal sulfides, metal nitrides, metal carbides, metal alloys, molecular sieves, etc., or they can be supported on other surfaces. Supported catalysts are popular in industrial applications because they are considered to be chemically, mechanically and thermally stable to a great extent. The supports can be organic such as polymers (*e.g.*, polystyrene), copolymers (*e.g.*, styrene-divinylbenzene) as well

DOI: 10.1201/9781003126270-7

as inorganic such as zeolites, alumina, activated carbon, titanium dioxide, silica and graphene [1].

Heterogeneous catalysts are mostly used in important organic reactions such as hydrogenation, enantioselective hydrogenation, hydrohalogenation, dehydrogenation, reduction of sulfides, reductive alkylation, coupling reactions (e.g., Heck reaction, Suzuki coupling) and hydrosilylation [2]. Most of these reactions are catalyzed by transition metals Pd, Pt, Ru, Rh, Ir and Re. For the last two decades, many new organic transformations are also reported where heterogeneous catalysts are used in different forms. The chapter will represent the recent developments in the field of organic synthesis where heterogeneous catalysts (supported metals and in pure form) are efficiently applied.

7.2 PALLADIUM-BASED HETEROGENEOUS CATALYSTS

Palladium has a unique place in organic synthesis especially in oxidation processes such as the Wacker process, alkene functionalization as well as C–C/C–H functionalization. Homogeneous Pd-based catalysts enable diverse C–C/C–H functionalization; however, their efficiency reduced due to the production of Pd-black which affected the activity of the catalysts. Another disadvantage of using Pd-salts as a catalyst, e.g., $Pd(OAc)_2$ in a homogeneous medium, is that recycling is very difficult especially when it is used with oxidants. To overcome this problem, Jan-E. Backvall and group have developed a Pd-based heterogeneous catalyst in which palladium nanoparticles were immobilized into the cavities of aminopropyl-functionalized siliceous mesocellular foam, denoted as Pd-AmP-MCF [3]. This catalyst is successfully used in various organic transformations such as oxidative carbocyclization-borylation of enallenols [4], synthesis of γ-lactones and γ-lactams [5] and furan and oxaborole derivatives [6] from enallenols through the oxidative tandem process.

The formation of various substituted cyclobutenol derivatives **2** is described in Figure 7.1 (representative examples are shown) through diastereoselective oxidative carbocyclization/borylation of enallenols **1** with heterogeneous Pd-catalyst and B_2pin_2 with excellent diastereoselectivity and yields up to 83%. The active catalyst in the reaction was Pd^{II}-AmP-MCF which is produced by the oxidative action of 1,4-benzoquinone (BQ) on Pd^0-AmP-MCF. Authors also explained the role of triethylamine (NEt_3) as an additive in the reaction that improved the yield of **2a** up to 82% and it is believed that NEt_3 interacted with Pd in the co-ordination process (the use of chiral trialkylamine, (3S)-N,N-dimethylpyrrolidin-3-amine produced cyclobutenol **2a** with poor enantiomeric excess, ee 16% and 62% yield) [4].

The reaction tolerated various functional groups such as methyl, butyl, benzyl, additional hydroxyl, esters and cycloalkylidenes (selected examples are included in Figure 7.1). It is worth noting that when a chiral enallenol (S)-**1a** (ee 95%) was used, the cyclobutenol (1S,4S)-**2a** was obtained without loss of any chirality. Additionally, the efficiency of the catalyst was tested seven times in rerun experiments and no activity loss or metal leaching was observed. The diastereoselectivity was not influenced even after several uses of catalyst.

The mechanism of transformation of enallenols to cyclobutenol was proposed on the basis of experimental analysis and deuterium experiments. It was observed

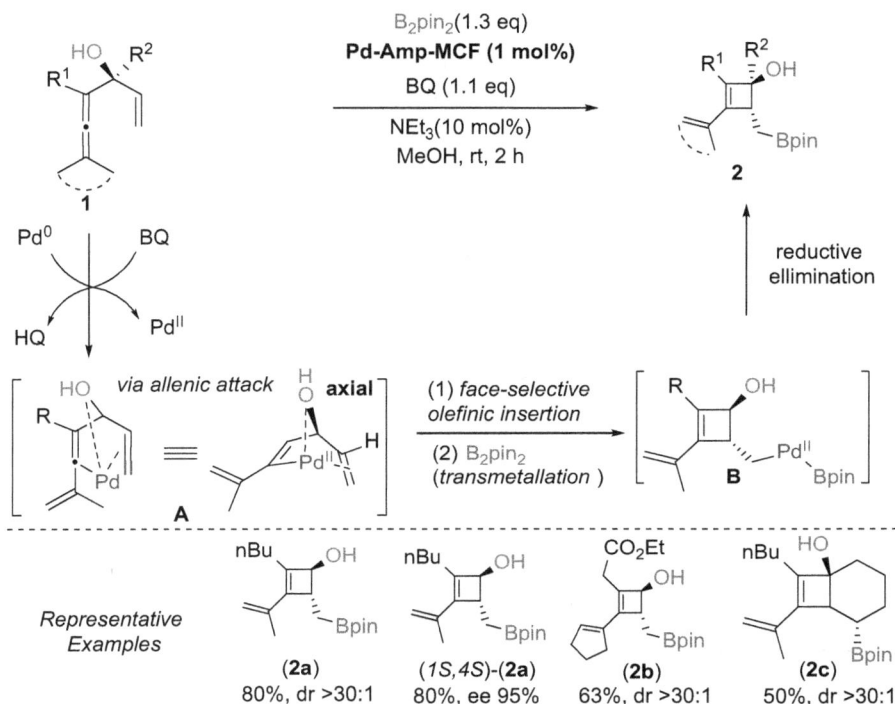

FIGURE 7.1   Diastereoselective oxidative carbocyclization/borylation of enallenols.

that the hydroxyl group (or -amine NH-R) was found to be essential for the diastereoselectivity of the reaction. Simultaneous coordination of allene, olefin and –OH group to the Pd^{II} center and then allenic attack would form intermediate **A** (Figure 7.1). Then, the hydroxyl group promotes the face-selective olefinic insertion and subsequent transmetalation would lead to intermediate **B** that furnishes product **2** after reductive elimination.

Jan-E. Backvall [5] further utilized the same Pd-catalyst for the chemodivergent and diastereoselective synthesis of γ-lactones and γ-lactams from enallenols **3** by oxidative carbocyclization of allenes in the presence of carbon monoxide (CO) where methyl-BQ was used as oxidant and CH_2Br_2 as a solvent at room temperature (Figure 7.2). The lactones produced had high diastereomeric ratios (observed > 50:1) with a minor amount of other lactones **5** as by-products (produced *via* intermediate **B** after lactonization, with ~4% maximum yield). The amine bases used as additives have shown no effect on the outcome of the reaction. The tandem carbonylation reaction of different enallenols **3** derivatives produced the library of lactones **4** with different groups when the scope of the reaction was investigated (key examples represented in Figure 7.2). The important characteristic of Pd^{II}-AmP-MCF is the recyclability which was tested ten times without loss in diastereoselectivity and catalyst activity for the compound **4a**. Furthermore, the strategy also provides an efficient method toward

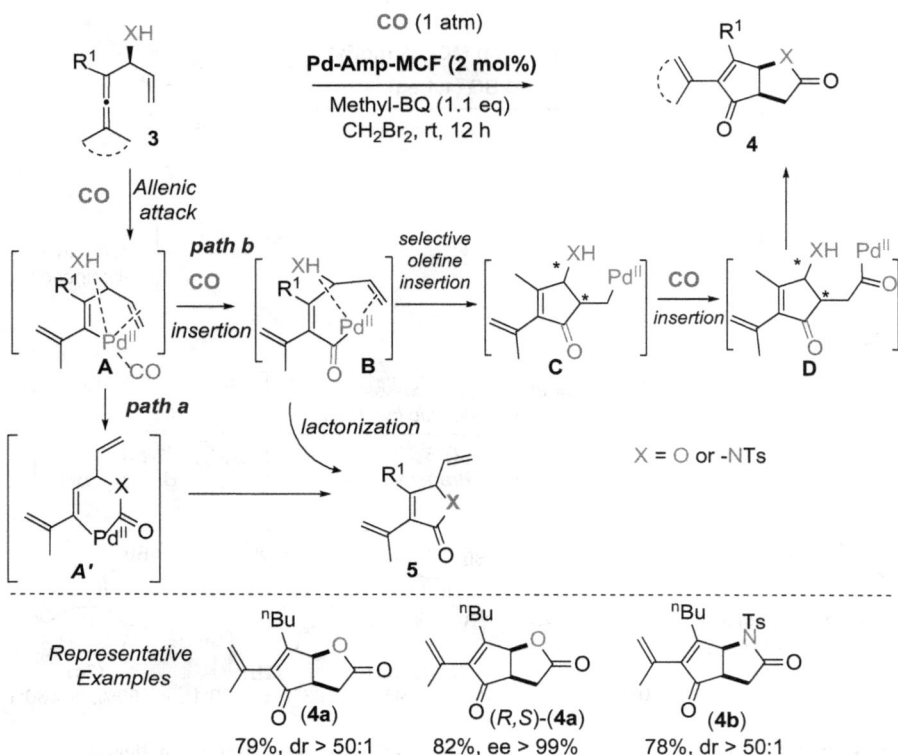

FIGURE 7.2 Oxidative carbocyclization of allenes.

chiral γ-lactones (*e.g.*, (*R,S*)-**4a**) with high step-economy, selectivity and enantiomeric excess (99% *ee*) if optically pure (*S*)-enallenols **3** was used as a substrate.

Interestingly, the reaction was chemoselective under different catalyst systems. In the homogeneous medium where Pd(OAc)$_2$ was used as a Pd^{II}-source, lactone **5** was the major product whereas heterogeneous Pd^{II}-AmP-MCF furnished lactone/lactam **4** as the main product. Mechanistically, simultaneous coordination of the hydroxyl group (or sulfonamide group) allene and olefin unit to the Pd^{II} center occurs at the initial step, which would first promote the allenic attack and then CO coordination to form common intermediate **A** (*Int*-**A**) in both catalyst systems (Figure 7.2). If homogeneous Pd-catalyst was used, attack by the XH group on coordinated CO would be favored with the formation of *Int*-**A'** and gives γ-lactones (or γ-lactams) **5** as the major product after reductive elimination (*path a*). However, the heterogeneous Pd-catalyst would favor CO insertion into the Pd—C bond of *Int*-**A** that would lead to *Int*-**B** (*path b*). Next, the selective olefin insertion occurs in *Int*-**B** directed by the OH (or NHTs) producing *Int*-**C** that undergo CO insertion affording *Int*-**D**, which follow lactonization (or lactamization) to give cyclopentenone-fused γ-lactones (or γ-lactams) **4**. It is also possible that lactonization or lactamization of *Int*-**B** could occur that lead to γ-lactones or γ-lactams with two chiral centers. In *Int*-**C**, alkyl-Pd and OH (or NHTs) groups are on the same side of cyclopentenone moiety explaining high diastereomeric ratios.

The enallenols are important substrates that provide a variety of organic compounds with recyclable heterogeneous catalyst Pd-AmP-MCF *via* tandem oxidative carbocyclization as we have seen in the above examples. In another recent report from the Jan-E. Backvall [6] where enallenols and alkynes as substrates are converted into highly substituted furan and oxaborole derivatives in a site- and stereoselective manner. In a standard reaction condition where catalyst Pd-AmP-MCF was used as a Pd^{II} source and BQ was used as an oxidant, the solvents played a major role in the outcomes of the reaction (Figure 7.3, Route A). If the rest of the reaction conditions are similar, the solvents CH$_3$CN, CH$_3$OH and CHCl$_3$ provided (Z)-tetrasubstituted olefin (**7**), 2,5-dihydrofuran (**8**) and tetrasubstituted furan (**9**) as the major products, respectively with excellent yields. Few representative examples of the scope of the reaction are shown in Figure 7.3 with each solvent. Moreover, the strategy was also successfully verified for the construction of oxaborole **10** with examples of chiral products (*e.g.*, *S*-**10b**, Route B, Figure 7.3) in the scope of the reaction.

Mesocellular silica foam (MCF) support is not only used for the preparation of metal-based catalysts and their use in various organic reactions but they are also used to prepare a heterogeneous "hybrid catalyst" system where Pd-nanoparticles and lipase enzyme *candida antarctica: lipase B* (CALB) were co-immobilized in the

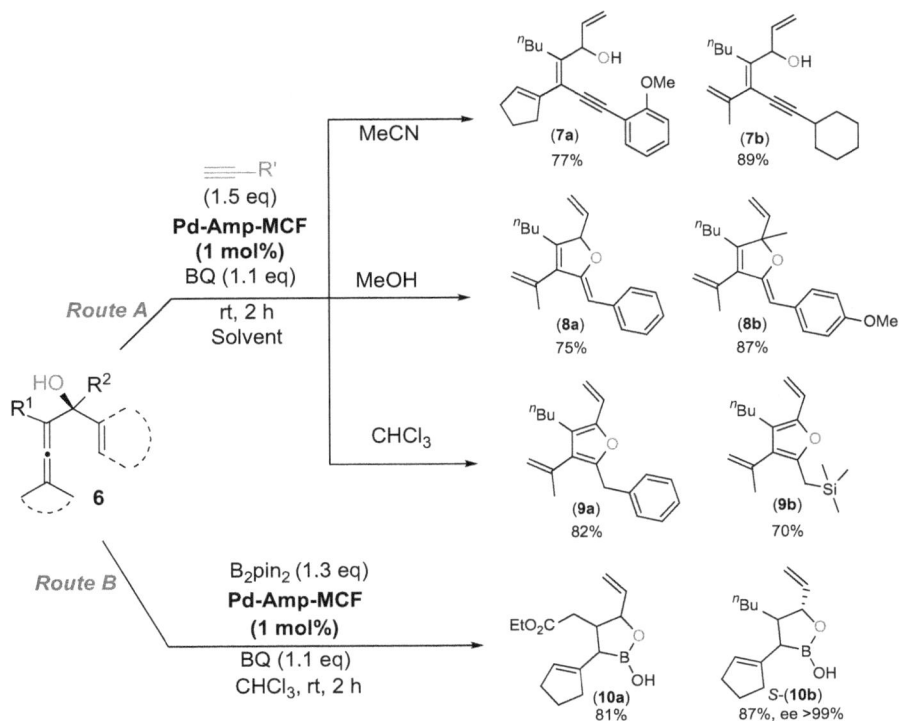

FIGURE 7.3   Synthesis of oxaboroles.

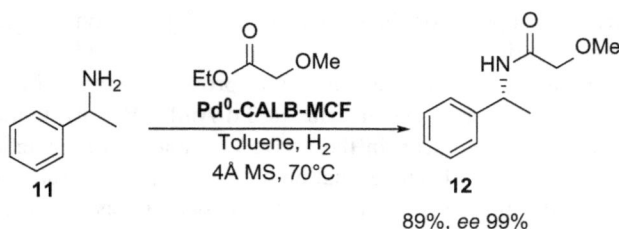

FIGURE 7.4 Deracemization of amines.

cavities of siliceous MCFs. The co-operative catalytic system was used for "*deracemization*" of amines where the lipase CALB acts as a resolving agent that selectively transforms only *R*-enantiomer of the given amine **11** to the corresponding amide **12** (Figure 7.4). The unreacted *S*-enantiomer of the amine **11** was racemized by the Pd-catalyst system that is further utilized by the enzyme catalyst CALB to continue until amine was completely consumed (Figure 7.4). The kinetic resolution of amines was highly efficient and produced enantiomerically pure amides with high yield and high enantiomeric ratio (*ee* > 99). In rerun experiments, the catalyst system was found to be efficient up to two runs; however, the reaction time had to be increased from 16 h to 48 h. In the third rerun experiment, the yield dropped even with longer experiment times though *ee* values remained high (>99%). The leaching tests of the metal showed low Pd-leaching and it was then proposed that protein denaturation of CALB was the reason for its deactivation which can be triggered by the polar surface of the MCFs.

Recently, W. Shi *et al.* [7] reported the preparation of a reusable Pd-based catalyst Pd-MOL-phenCl **13a** (Figure 7.5), constructed on metal–organic layers (MOLs) which become an active species as F-Pd-MOL-phenCl **13b** for fluorination of arenes by treating with selectfluor. This work has shown that MOLs and MOFs can be used to stabilize unique active catalyst sites and solid reagents/catalysts can be made for late-stage functionalization. The catalyst **13a** has shown different regioselectivities as compared to its Pd-based solid reagent, [F-Pd-MOL-phenCl], however, it has shown similar outcomes in terms of yields and regioselectivity with homogeneous catalyst **13c**.

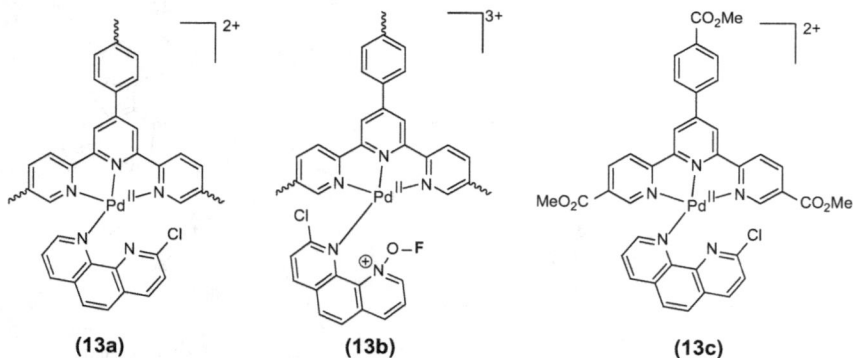

FIGURE 7.5 Heterogeneous catalyst, Pd-MOL-phenCl (**13a**), active catalyst F-Pd-MOL-phenCl (**13b**) and homogeneous catalyst, [Pd^{II}(TPY-OMe)(phenCl)]$^{2+}$ (**13c**).

TABLE 7.1

Fluorination of Arenes by Pd-MOL-phenCl

Conditions
- (a) *Heterogeneous catalyst*:
  13a (4 mol%), Selectfluor (1–2 equiv.), MeCN, 25–50°C, 36 h
- (b) *Homogeneous catalyst*:
  13c (4 mol%)+ Selectfluor (2 equiv.), MeCN, 50°C, 36 h

S. No.	Substrate	Heterogeneous Catalyst 13a		Homogeneous Catalyst 13c	
		Yield (%)	Regioselectivity (*ortho:para*)	Yield (%)	Regioselectivity (*ortho:para*)
1.	Toluene	55	45:55	31	57:43
2.	Phenol	76	25:22[a]	75	41:44[a]
3.	[1,1′-Biphenyl]-4-carbonitrile	90	58:42	97	76:24
4.	2-Phenylpyridine	82	56:41	54	67:33
5.	4-Phenylcyclohexan-1-one	62	51:49	95	35:28[a]

[a] Multi-fluorinated products.

The scope of the reaction with catalyst **13a** is shown in Table 7.1 with selected examples in comparison with homogeneous catalyst **13c**. Fluorination of toluene as a substrate produced 55% yield with catalyst **13a** with regioselectivity 45:55 (*ortho* to *para* fluorination), whereas its homogeneous counterpart produced a lower yield of 31% with a slightly better regioselectivity of 57:43. The fluorination of phenol gave regioselectivity of around 1:1 with similar yields with both catalysts (entry 2). However, both catalysts produced multi-fluorinated products along with *ortho*- and *para*-isomers. In entries 3 and 4, the yields were better in the case of catalyst **13a** but the homogeneous catalyst **13c** showed better regioselectivity.

Further, upon fluorination of 4-phenylcyclohexanone (entry 5), the catalyst **13a** only produced *ortho* and *para* isomers, whereas in the case of catalyst **13c**, fluorination of the aliphatic C–H bonds were also observed along with *ortho* and *para* isomers (structure of the products not shown). Table 7.1 concludes that the heterogeneous catalyst has similar results with homogeneous catalysts. However, one advantage of using heterogeneous catalyst **13a** over **13c** is that **13a** is recyclable and its catalytic activity is not diminished even after three consecutive usages. The authors reported only negligible (0.01–0.04%) Pd-leaching of the total Pd in the catalyst.

One of the limitations of hydrogenation is that it requires high temperature and high pressure of H_2 gas. To overcome this difficulty, highly active platinum, palladium

$$\text{alkene or alkyne} \xrightarrow[\text{H}_2\text{O, rt (8-15 min)}]{\overset{\text{NaBH}_4}{\text{Pd/SPC}}} \text{alkane}$$

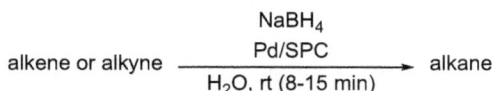

FIGURE 7.6   Reduction of alkene/alkynes with NaBH$_4$ and Pd/SPC.

and nickel catalysts are obtained by reduction of metal salts with sodium borohydride (NaBH$_4$). This makes NaBH$_4$ a suitable alternative for hydrogen gas and can be used for reduction or hydrogenation. Besides this, NaBH$_4$ is a cost-effective and easily available metal hydride that has many advantages like easy storage, can be used in different solvents and is safe to use. It provides chemoselectivity, regioselectivity and diastereoselectivity during reduction. Moreover, when mixed with simple metal salts it provides selective reduction of organic compounds. For instance, when NaBH$_4$ is combined with simple salts of palladium, rhodium or nickel, it does the selective reduction of alkenes and alkynes. Figure 7.6 shows the reaction of palladium metal supported on sulfonated porous carbon (SPC) for the reduction of alkenes and alkynes to the corresponding saturated alkanes by sodium borohydride. In this reaction, no external source for hydrogen gas is needed as NaBH$_4$ in an aqueous solution releases hydrogen [8].

Lindlar's catalyst is an example of a heterogeneous catalyst which is composed of palladium metal deposited on calcium carbonate or barium sulfate and poisoned with quinoline or lead acetate. The "poisoning" is done to reduce the normal catalytic activity of the catalyst in order to stop hydrogenation after the reduction of the first π-bond of alkynes. If poisoning is not done, the reduction will lead to fully hydrogenated alkanes.

Lindlar's catalyst requires the careful optimization of reaction conditions (temperature and pressure) because even minor changes can lead to undesired products of the hydrogenation of alkynes [9]. Selective hydrogenation of substituted alkynes (controlled or partial hydrogenation to half hydrogenated alkenes) is usually carried out in the presence of Lindlar's catalyst which is also a stereoselective catalyst for the reduction of substituted alkynes to *cis*-alkenes (Z-alkenes) as described in Figure 7.7 [10].

7.3   PLATINUM-BASED HETEROGENEOUS CATALYSTS

The catalytic hydrogenation of olefinic bonds is a key reaction in chemistry that has a broad scope of industrial applications such as synthesis of fine chemicals and food chemistry. The π-bond of an alkene or alkyne is relatively weaker than the sigma (σ) bond and can be broken easily to add reagents. Transition metals such as platinum, palladium, rhodium, ruthenium and nickel are commonly used as heterogeneous catalysts for reductions of unsaturated bonds in organic compounds (Figure 7.8). The metals are used in various forms for the reduction of unsaturated bonds. Some examples of metal-based catalysts are palladium in the form Pd–C, platinum in the form PtO$_2$ and nickel in the form Raney-nickel (Ra-Ni). Mostly, platinum or palladium is used in a finely divided state adsorbed on charcoal (Pt/C or Pd/C) or alumina, and the hydrogenation reaction takes place in an atmosphere of hydrogen gas. Metal in a finely divided state has a large surface area and hence it is more active.

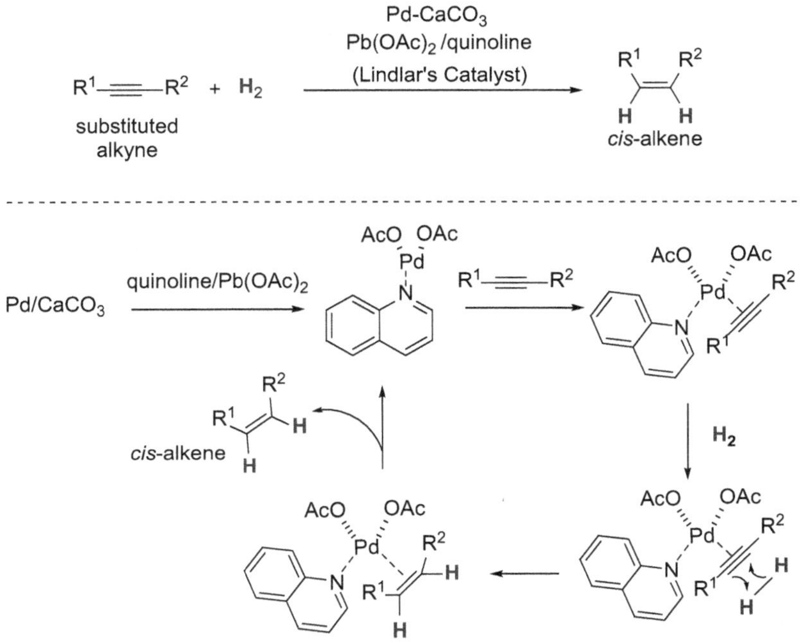

FIGURE 7.7   Hydrogenation of substituted alkynes with Lindlar's catalyst.

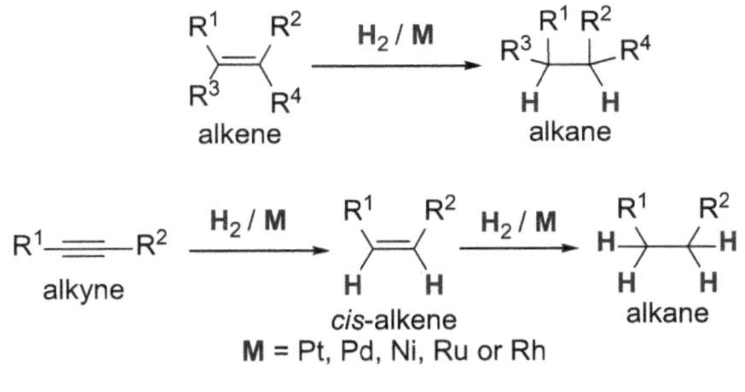

FIGURE 7.8   Catalytic hydrogenation of unsaturated bonds.

Hydrogenation of an unsaturated bond is a thermodynamically favorable reaction because it forms a more stable alkane product. Even though the hydrogenation of an alkene is thermodynamically favored, it does not proceed without a proper catalyst. The addition of a catalyst enables a hydrogenation reaction to occur by lowering the activation energy of reactants and it helps to reach the transition state easily to form the product.

First, both the reactants, hydrogen and ethene are chemisorbed on the surface of the metal catalyst. The first C–H bond formation step occurs from a C-atom bonded to Pt in a classical three-center transition state (C, Pt, H) [11]. The process is further repeated for the addition of the second H-atom to the remaining C-atom of ethene

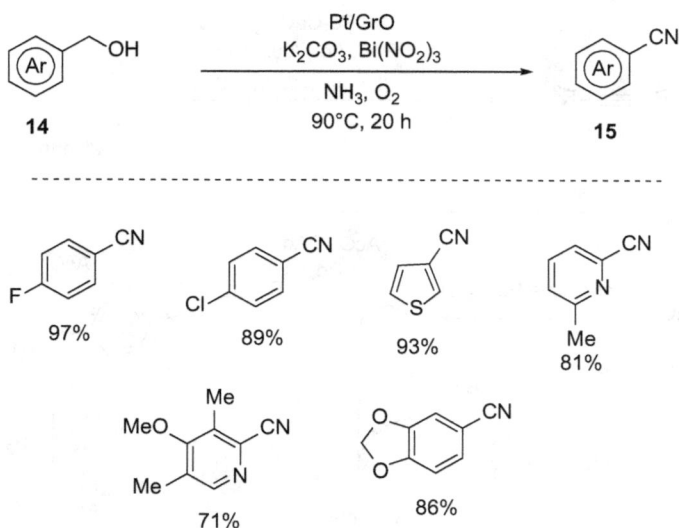

FIGURE 7.9 Nitrile synthesis by Pt/GO heterogeneous catalyst.

resulting in the formation of ethane which has a low affinity for the metal surface and, thus easily desorbed, creating a vacant space for the adsorption of new ethene and hydrogen molecules. The addition of hydrogen to an olefinic bond is a *syn*-addition process attributed to the physical arrangement of the alkene and the hydrogens on a flat metal catalyst surface which describes that the two hydrogens must be added to the same face of the double bond.

Apart from hydrogenation, Pt-based heterogeneous catalysts have been widely used in the oxidation of alcohols to aldehydes and carboxylic acids. Recently, S. S. Stahl *et al.* [12] reported that Pt-catalyst based on graphene-oxide (GrO) can be used for the synthesis of nitrile **15** derivatives through the oxidative coupling of alcohols **14** and ammonia under aerobic conditions. With the combination of Pt–GrO catalyst with K_2CO_3 and $Bi(NO_2)_3$ as additives, the optimized reaction conditions effectively produced various aryl and aryl(hetero) nitrile derivatives in the presence of ammonia and oxygen in excellent yields (Figure 7.9).

7.4   RUTHENIUM-BASED HETEROGENEOUS CATALYSTS

The group of K. Wada [13] reported a simple Ru/CeO_2 heterogeneous catalyst system that was efficiently used for transfer-allylation from homoallyl alcohols **16** to aldehydes **17** producing saturated ketones **18** in high yields (Figure 7.10).

The catalytic system was recovered and calcinated at high temperature (400 °C for 30 minutes), and then reused up to three times without observing any loss of catalytic activity. K. Wada further modified the catalyst Ru/CeO_2 in the presence of PPh_3 denoted as $3PPh_3$-Ru/CeO_2 and this heterogeneous catalyst was employed for the direct C–H arylation of aromatic compounds [14]. The authors claimed for the first time the use of such a catalysis system where nitrogen-directed arylation of aromatic

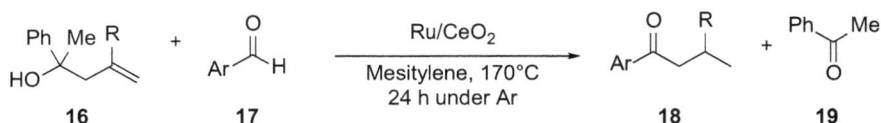

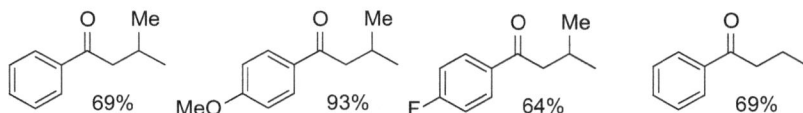

FIGURE 7.10 Transfer-allylation from homoallyl alcohols to aldehydes.

C–H bonds occurred with aryl halides. Under optimized reaction conditions, the reaction of benzo[h]quinoline **20** and aryl halides **21** in N-methyl-2-pyrrolidone (NMP) at 140 °C in the presence of 3PPh$_3$-Ru/CeO$_2$ afforded various arylated products **22** in good to excellent yields. The catalyst could be recovered by centrifugation, then washed with organic solvent/H$_2$O and calcinated at 400 °C before remodification with PPh$_3$ in the presence of H$_2$-atmosphere to get the recycled catalyst 3PPh$_3$-Ru/CeO$_2$ again for reuse (Figure 7.11). The authors performed recycling experiments three times without any loss of catalytic activity.

L. Ackermann *et al.* [15] reported a ruthenium-based silica-derived heterogeneous catalyst, Ru@SiO$_2$ that was successfully used for *meta*-C–H functionalization of

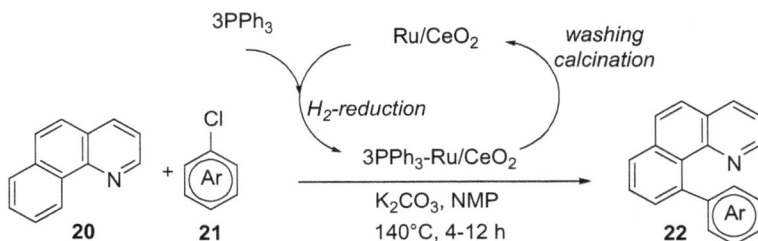

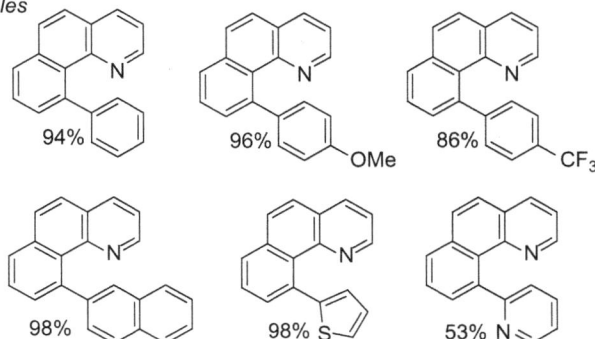

FIGURE 7.11 C–H arylation of aromatic compounds.

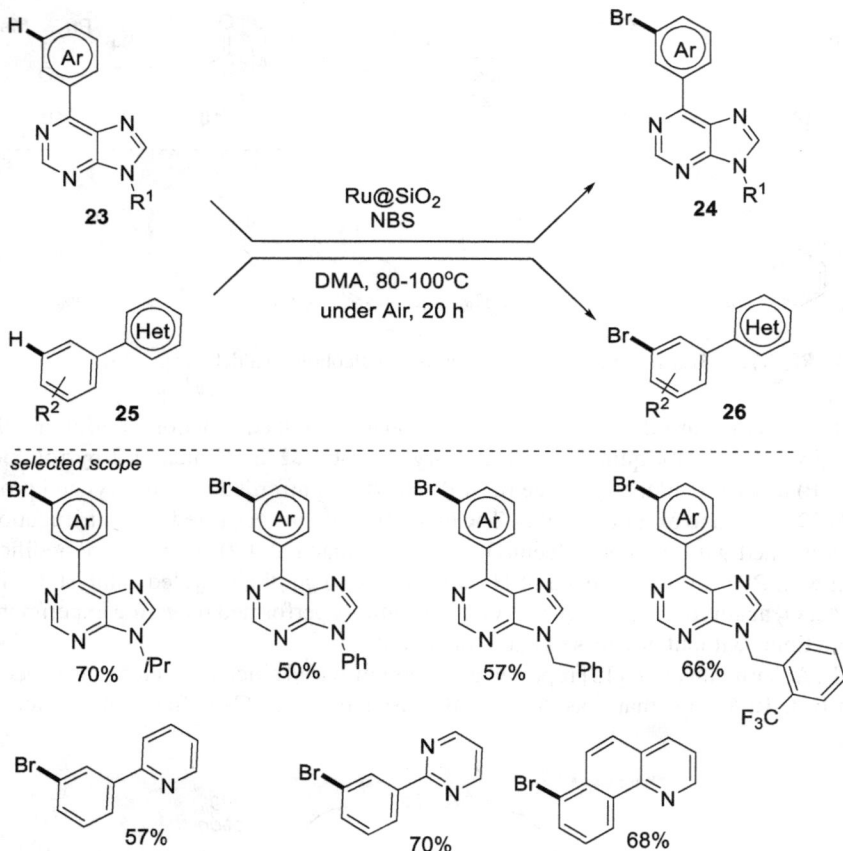

FIGURE 7.12 *meta*-C–H functionalization of purine derivatives.

purine derivatives **23** and substituted arenes **25** (Figure 7.12). The reaction produced a broad scope of selective *meta*-brominated products **24** and **26** with high yields and the catalyst Ru@SiO$_2$ was also found to be uniquely selective for *meta*-C–H functionalization in comparison to other homogeneous metal catalysts based on palladium, rhodium, iridium or cobalt complexes where the latter produced non-selective mono- or multi-bromination products. Figure 7.12 described the reaction scope for purine nucleobases derivatives and substituted arenes. The recyclability of the heterogeneous catalyst was tested by rerun experiments and it was found that the catalyst was consistent in activity and selectivity of meta-bromination of purine was up to seven runs with a slight decrease in the reaction yields in the sixth and seventh runs.

7.5 TITANIUM-BASED HETEROGENEOUS CATALYSTS

Oxidation reaction is considered an important reaction in chemical industries that produce a wide range of applications. For example, oxidation of ethylbenzene produces key products such as acetophenone and 1-phenylethanol that are used as precursors for the synthesis of a variety of drugs, chiral alcohols, hydrazones and chalcones [16]. For oxidation reactions, homogeneous catalysts have been extensively used for the synthesis of bulk as well as fine chemicals. However, homogeneous catalysts are non-recyclable, environmentally hazardous and also, they corrode the industrial materials raising the maintenance costs. To sort out these problems, the homogeneous catalysts could be prepared by the dispersion of a metal on an insoluble solid support to keep the metal on the surface where the catalysis reaction takes place [17]. Therefore, heterogeneous catalysts are considered to be a better choice for organic synthesis because they are easy to separate at the end of the reaction and also, they are easy to handle.

The common oxidants that are used as an oxygen source in oxidation reactions are TBHP (*tert*-BuO$_2$H) and hydrogen peroxide (H$_2$O$_2$), which are generally employed with a heterogeneous catalyst system such as titanium silicate catalysts. Titanium silicalites (TS-1 and TS-2) are microporous solid materials made of SiO$_2$ and TiO$_2$ that have silicalite structures. These catalysts have shown good activity, selectivity and stability below 100 °C. TS-1 is a popular catalyst for various types of organic reactions especially oxidation reactions that include conversion of olefins to epoxides, arenes to phenol derivatives, ketones to oximes, primary alcohols to aldehydes and acids, secondary alcohols to ketones and alkanes to secondary and tertiary alcohols/ketones. The general scope of oxidation reactions with TS-1/HP is shown in Figure 7.13. The remarkable catalytic activity of TS-1 in these reactions is due to the site isolation of Ti(IV) centers in the hydrophobic pores of silicalites that permits simultaneous adsorption of the hydrophobic substrate and the oxidant.

The ring hydroxylation of phenol with TS-1/H$_2$O$_2$ produces a ring hydroxylated product. However, in a heteroaromatic ring such as furan, a cyclic, dienic ether which

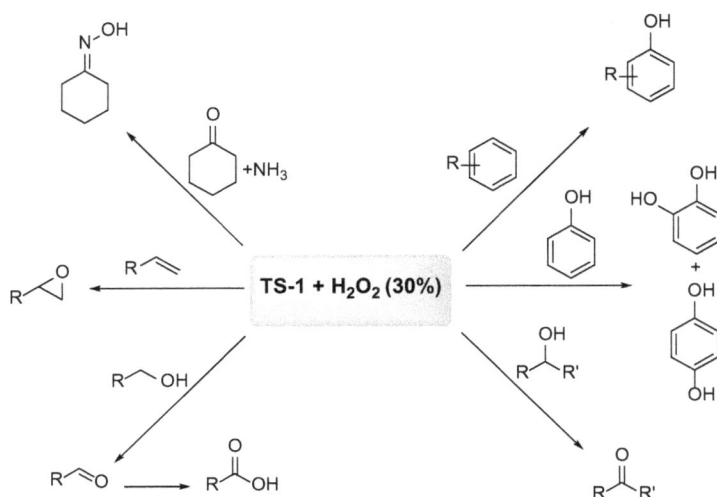

FIGURE 7.13 General scope of oxidation reactions with TS-1/HP.

is stabilized by resonance, different reaction pathways are possible in its oxidation with TS-1/H_2O_2. J. Wahlen *et al.* [18] reported the oxidation of furan derivatives with TS-1/H_2O_2 (Si:Ti = 35:1) produced 2-butene-1,4-dial (**29/30**) instead of 2-hydroxy-furan (Figure 7.14(a)). In the given reaction sequence, the plausible path of the formation of dialdehydes is through the formation of unstable epoxide **28** on one of the double bonds of furan **27** that would rearrange to corresponding dialdehydes **29** and **30**. The reaction was highly selective and the observed ratio of *cis* to *trans* dialdehyde products was 94:6. Further, furfuryl alcohol **31** was also oxidized under similar conditions (Figure 7.14(b)). Oxidation of furfuryl alcohol with TS-1/H_2O_2 in acetonitrile gave quantitative conversion to 6-hydroxy-2H-pyran-3(6H)-one **33** in 90% yield after 24 h reaction. The mechanism of the reaction was also explained on the basis of epoxide formation and then its rearrangement to an enedione alcohol **32** which undergoes intramolecular cyclization by the attack of the hydroxy group on the aldehyde (Figure 7.14(b)). The reaction furnished pyran derivative **33** with good selectivity (93%) and conversion (>99%). The catalyst TS-1 was tested for its reusability up to three times where its activity was reduced over each cycle with possible blockage of active sites by the decomposition of the products. However, the catalytic activity of TS-1 was fully restored upon calcination.

In a recent report by X. Liu *et al.* [19], the oxidation of 1-alkene was tested with TS-1/H_2O_2 catalyst system. The project aimed to investigate the formation of trace amounts of dimethoxymethane as a by-product during the oxidation of propylene to propylene oxide (HPPO process) where H_2O_2 is used as an oxidizing agent. The dimethoxymethane is produced by the aldol condensation between methanol and formaldehyde produced during the oxidation process of propene. Using 1-hexene as a probe, the investigation resulted that using ethyl acetate as a solvent under given conditions, the reaction produced *n*-valeraldehyde as a prominent product along with corresponding epoxide and other undesirables (R = *n*-C_4H_9, Figure 7.15). However, the outcome of the reaction could be controlled by the addition of additives ($Na_2HPO_4 \cdot 12H_2O$ and NaH_2PO_4). In the presence of these additives, the formation of aldehyde was considerably suppressed.

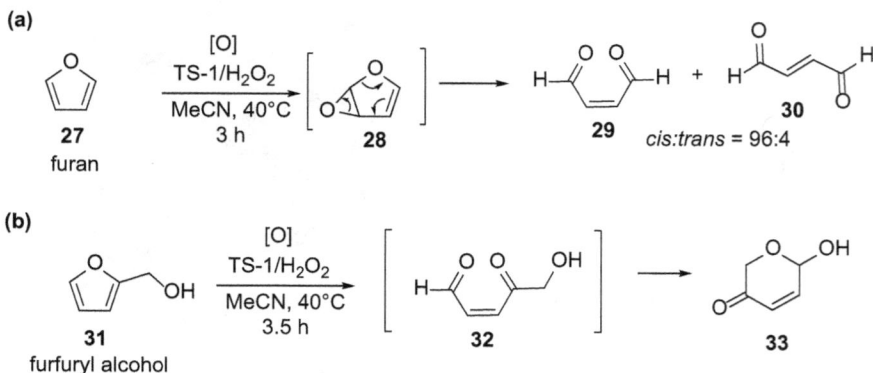

FIGURE 7.14 Ring hydroxylation of furan derivatives with TS-1/H_2O_2.

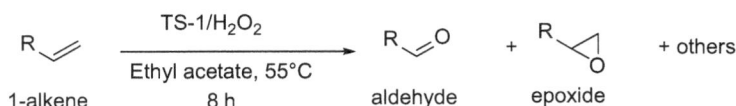

FIGURE 7.15 Oxidation of 1-alkenes with TS-1/H_2O_2.

With the optimized conditions for the better formation of aldehyde with 1-hexene, they also performed oxidation reactions with other alkene substrates (Table 7.2). The highest conversion was shown by 1-pentene with 65% selectivity, whereas 1-heptene showed the lowest conversion and selectivity because of the limited conversion due to the pore size of TS-1. In the case of 1-styrene, the aldehyde was the main oxidation product because of its higher electron density on the C=C bond.

Mechanistically, the formation of aldehyde was predicted through the addition and oxidative cleavage of C=C bond of 1-alkene in an acidic environment (as shown in Figure 7.16). When TS-1 comes in contact with H_2O_2, Ti^{IV}-OOH **B** is generated which forms an unstable intermediate **C** with the addition of a double bond that produced lower aldehydes upon hydrolysis. The authors did not explain the role of ethyl acetate in the production of aldehyde; however, they explained the role of additives, $Na_2HPO_4 \cdot 12H_2O$ and NaH_2PO_4 that could reduce the acidity of weak acid center on the TS-1 surface.

The catalyst system TS-1/H_2O_2 was also used by Y. Rodenas et al. [20] to oxidize furfural, obtained from renewable sources to convert it to an industrially important compound, maleic acid. The biomass-derived furfural is commercially available and can be obtained from lignocellulosic agro residues like corn cobs or sugar cane. The conversion of furfural **34** to maleic acid **36** proceeds via the removal of a C-atom in the form of formic acid producing 5-hydroxy-furan-2(5H)-one **35** which is further oxidized to maleic acid **36** along with other minor oxidized products such as fumaric acid, tartaric acid and malic acid (Figure 7.17). The reaction was quite efficient and utilized three moles of oxidant H_2O_2 per mole of furfural (high-grade) achieving maleic acid yields of more than 70% depending upon the solvent system used. While using a monophasic aqueous solution, though the reaction rate was faster and yields were observed up to 65%, the catalyst TS-1 suffered a mild deactivation possibly

TABLE 7.2
Oxidation of 1-alkenes with TS-1/H_2O_2

1-Alkene	Conversion (%)	Selectivity of Products (%)		
		Aldehyde	Epoxide	Other Products
1-Pentene	64	65	2	33
1-Hexene	59	51	27	22
1-Heptene	20	37	51	12
1-Styrene	54	90	<1	9

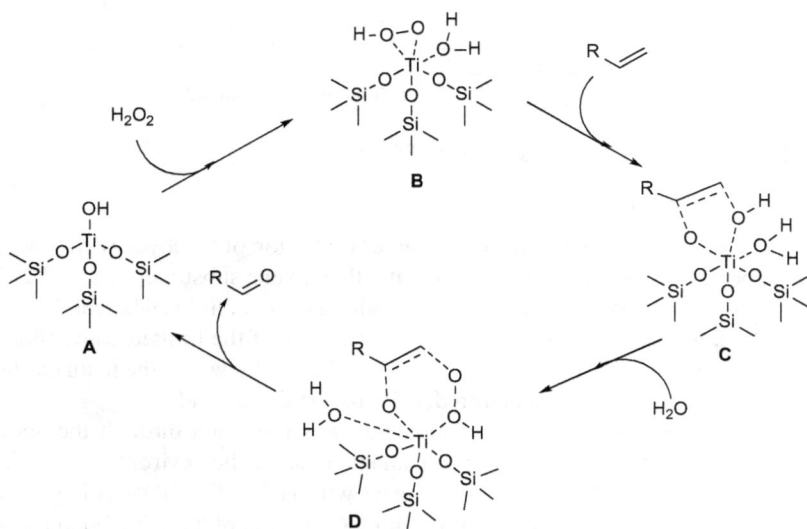

FIGURE 7.16   The possible mechanism of oxidation of 1-hexene to *n*-valeraldehyde.

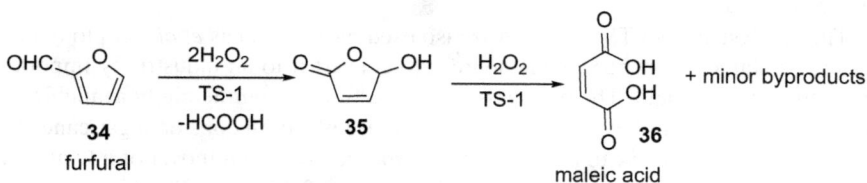

FIGURE 7.17   Oxidation of furfural by TS-1/H$_2$O$_2$.

because of the deposition of insoluble by-products on the surface or/and within the cavities. However, when the reaction was carried out in a mixture of solvents γ-valerolactone (GVL):water (from ratio 10:90 to 30:70), though the reaction proceeded slowly, it produced high yields of more than 70% and the catalyst could be reutilized up to 21 runs. Although the yields dropped around 40% in rerun experiments, it remained consistent after the catalyst system was stabilized after the 11th run. Further, the catalyst activity could be restored by the calcination in air at 550 °C. In their further studies, the yields could be further improved up to 83% using a mesoporous TS-1 catalyst for high-grade furfural and 73% for low-grade furfural (obtained from biomass). The co-solvent GVL prevented the deposition of by-products over the TS-1 surface or within the cavities of the catalyst. This improved the stability of the catalyst and also, GVL made separation of maleic acid easier by precipitation in the form of monosodium maleate.

Titanium oxide (TiO$_2$) is a cheap, environmentally acceptable and low-cost material that can be used as a photocatalyst in various photochemical reactions. C. Vila and M. Rueping [21] reported that the recyclable TiO$_2$ can be used as a photocatalyst for an oxidative Ugi-type three component multicomponent reaction (3C-MCR), where *tert*-amines can be converted to α-amino acids through C–H functionalization of *tert*-amines. In this reaction, visible light was used as a drive for organic transformation (Figure 7.18).

The best reaction conditions were obtained where different amines **37**, water and isocyanide derivatives **38** were reacted in dioxane as a solvent in the presence of TiO_2 under an 11-W lamp as a visible light source. The reaction provided α-amino acids **39** in moderate to good yields (Figure 7.18, plausible mechanism and selected scope of the reaction is represented). Moreover, the TiO_2 could be recovered and recycled for five times without loss in the catalytic reactivity and selectivity, and yields remained consistent in all rerun experiments.

The same group of M. Rueping [22] also reported a heterogeneous catalyzed C–H arylation of heteroarenes in the presence of visible light where TiO_2 played a dual role in the catalytic process: formation of TiO_2-azoether **A** and its initiation for their further conversion (Figure 7.19). Under optimal reaction conditions, various heteroarenes **41** (furan, thiophene and pyridine were used for direct C–H arylation) and different aryldiazonium salts **40** in EtOH were reacted in the presence of TiO_2 and visible light lamp at room temperature (Figure 7.19). The reaction represented a good reaction scope with different aryldiazonium salts affording coupling product in good yields (selected examples are shown in the scheme along with possible mechanistic route). The catalyst could be further recycled for five consecutive runs and no activity or selectivity loss was observed.

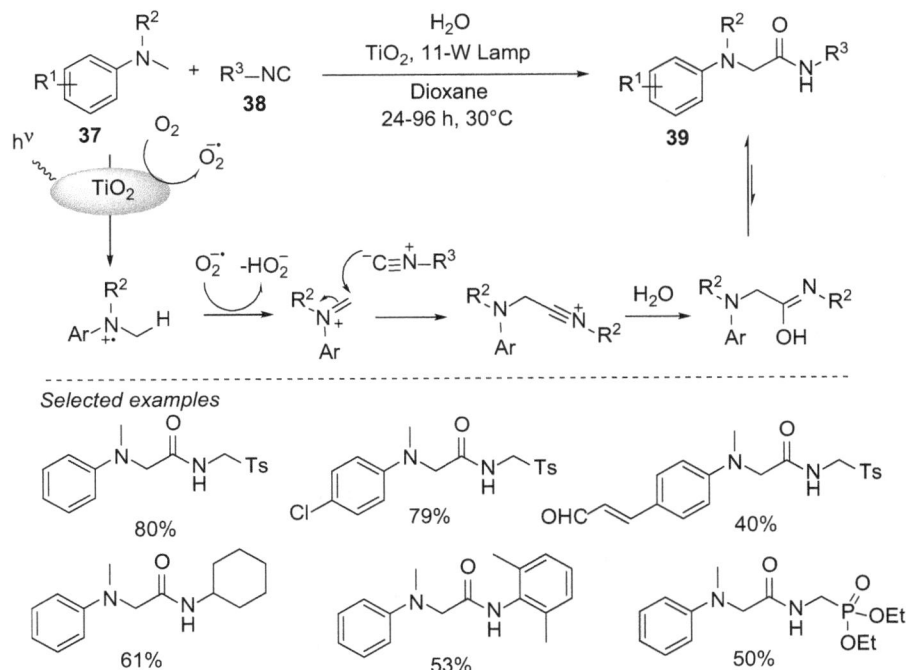

FIGURE 7.18 Ugi-type three component multicomponent reaction.

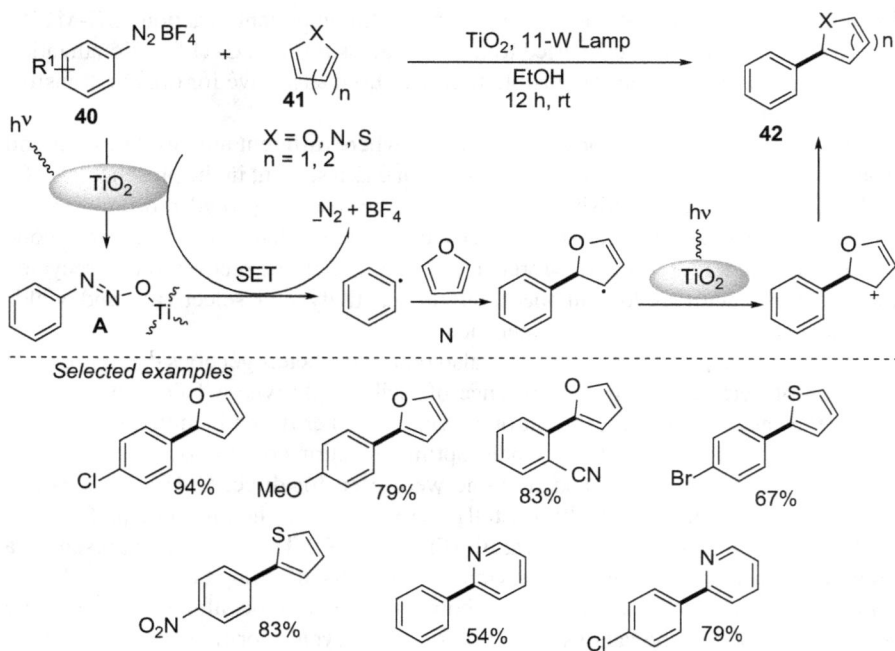

FIGURE 7.19 C–H arylation of heteroarenes.

7.6   RHODIUM-BASED HETEROGENEOUS CATALYSTS

Hydroformylation of olefins is one of the most important reactions in chemical industries. Wilkinson's catalyst is mostly used for hydroformylation of short-chain olefins. Hydroformylation reaction with Rh-based catalysts can be performed at lower
pressure in comparison to other catalysts. However, there are disadvantages associated with Rh homogeneous catalysts, so to keep their properties intact, they are
usually heterogenized by various methods such as: Rh on solid support, metallic Rh
nanoparticles on solid support, single-atom Rh on solid support and Rh-complexes on
supported liquids [23]. J. Amsler *et al.* [24] tested the potential of Rh-based oxide-
supported heterogeneous single-atom catalyst (SAC) for hydroformylation. They
investigated the heterogeneous hydroformylation both theoretically and experimen-
tally and found that the activity of the catalyst Rh-SAC dispersed on CeO_2 (Rh_1/CeO_2)
surface is similar to the molecular catalysts such as $HRh(CO)_4$ and $HRh(CO)_3PMe_3$.
In comparative analysis with Rh_1/CeO_2, they have also prepared Rh-SAC on MgO
(Rh_1/MgO), however, the latter catalyst being more stable, showed no catalytic activity.

Table 7.3 summarized the result of hydroformylation reaction of styrene with
Rh-SAC. The catalyst showed good activity with styrene **43** conversion to aldehyde
44 up to 96% in 12 h at 363 K (entry 4). This result was comparable to the molecular
catalyst $HRh(PPh_3)_3CO$ under similar reaction conditions (entry 6). In comparison to
Rh/C solid catalyst, Rh_1/CeO_2 was even better and showed higher TOF (over five
times more) with even less formation of undesired ethyl benzene **45** by olefin

TABLE 7.3
Hydroformylation of Styrene by Rh_1/CeO_2

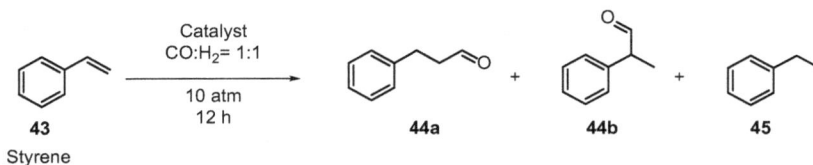

Entry	Catalyst[a]	Conversion (%)	Temperature (K)	TOF[b] (h^{-1})	44 (44a + 44b)	45
					Selectivity (%)	
1	Rh_1/CeO_2	42	333	25	98(31 + 67)	2
2	Rh_1/CeO_2	63	343	48	98(36 + 62)	2
3	Rh_1/CeO_2	93	353	88	97(45 + 52)	3
4	Rh_1/CeO_2	96	363	130	97(51 + 46)	2[c]
5	Rh/C (5 mol%)	85	363	24	85(47 + 38)	8[c]
6	$HRh(PPh_3)_3CO$	98	363	117	98(52 + 46)	0[c]

[a] Molar ratio of rhodium to substrate of 1:500.
[b] Turnover frequency: *It is the number of molecules of olefin substrate converted per total Rh atom per hour.*
[c] Other by-products.

hydrogenation (2% vs. 8%, entry 5). The regioselectivity of the reaction was found to be temperature-dependent. The ratio of linear to branched hydroformylation (**44a** vs **44b**) was increased with an increase in temperature (entries 1–5) which is found similar to the molecular catalyst in solution. The combined study proved that oxide-supported Rh-SAC, Rh_1/CeO_2 was highly active for hydroformylation of olefins and its performance as a heterogeneous catalyst was comparable to an $HRh(PPh_3)_3CO$ molecular catalyst.

The power of Rh-SAC was also tested by T. Zhang *et al.* [25] where rhodium single atom supported on ZnO nanowires (ZnO-nw) was used for hydroformylation of different olefins **46** to aldehyde products **47** with similar efficacy and activity as traditional homogeneous Wilkinson's catalyst, $RhCl(PPh_3)$ (Table 7.4). The author claimed that the SAC, Rh_1/ZnO-nw was found to be more active (in one or two order of magnitude) as compared to the previously reported heterogeneous catalysts. Also, in hydroformylation reaction of olefins, Rh_1/ZnO-nw showed 99% chemoselectivity and hydrogenation product (*e.g.*, ethylbenzene in case of styrene as a substrate) was not observed. Moreover, complete conversion of olefin was observed with very low catalyst loading (0.006%).

Table 7.4 represents the scope of hydroformylation reaction with Rh_1/ZnO-nw as catalyst with different styrene derivatives and aliphatic olefins **46**. The study was also compared with homogeneous catalyst $RhCl(PPh_3)$ and as shown in Table 7.4, hydroformylation resulted in similar outcomes as traditional catalysts,

TABLE 7.4
Hydroformylation of Olefins by Rh$_1$/ZnO-nw

$$R\diagup\!\!=\!\!\diagdown \quad\xrightarrow[\text{toulene, 100°C, 12 h}]{\substack{\text{CO:H}_2\ (0.8\ \text{MPa: 0.8 MPa})\\ \text{Rh}_1/\text{ZnO-nw}\ (0.006\%)}}\quad R\diagup\!\!\diagdown^{\text{CHO}}$$

46 → **47**

S. No.	Substrate	Selectivity (%)		Turnover Number (TON)	
		Rh$_1$/ZnO-nw	RhCl(PPh$_3$)	Rh$_1$/ZnO-nw	RhCl(PPh$_3$)
1	Styrene	99	92	40,000	19,000
2	4-Methylstyrene	99	—	26,000	—
3	4-Chlorostyrene	99	90	45,000	20,000
4	Prop-1-en-2-ylbenzene	99	—	4000	—
5	1-Hexene	86	—	10,000	—
6	Propene	99	99	15,000	37,000

with an even better turnover number (TON) in most of the cases (highest in the case of *para*-chlorostyrene). Additional advantage of using Rh$_1$/ZnO-nw was its reusability as the catalyst was recycled up to four rerun experiments without any significant loss in its activity and selectivity and no metal leaching was observed (Figure 7.20).

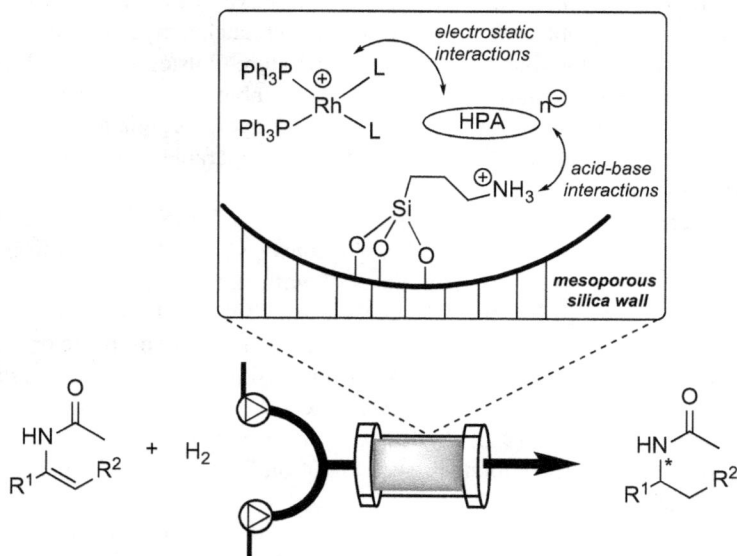

FIGURE 7.20　Rh-based heterogeneous chiral catalyst-mediated asymmetric hydrogenation of enamides.

Y. Saito and S. Kobayashi [26] reported a robust Rh-based heterogeneous chiral catalyst system that utilized acid–base and electrostatic interactions for asymmetric hydrogenation of enamides and α-/β-dehydroamino acids in a continuous flow process. The Rh-catalyst was developed based on the hypothesis that strong interactions are possible between heteropolyacids (HPA, used for immobilization of catalytic species on a solid support *via* ionic interactions) and a support having Brønsted basicity on its surface. It was expected that acid–base interactions hold HPA tightly to the support (salt-formation) that will prevent active metal leaching and also HPA has multivalent anions that will form ion pairs with cationic metal species. The chiral Rh catalysts were immobilized on an HPA/amine-functionalized mesoporous silica composite (Figure 7.21).

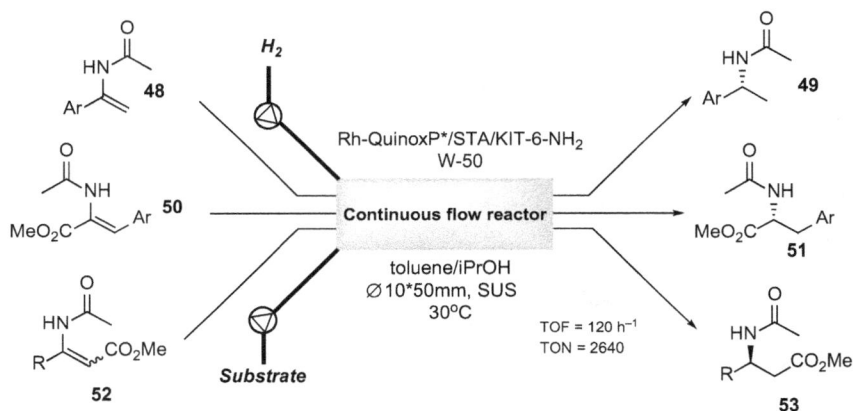

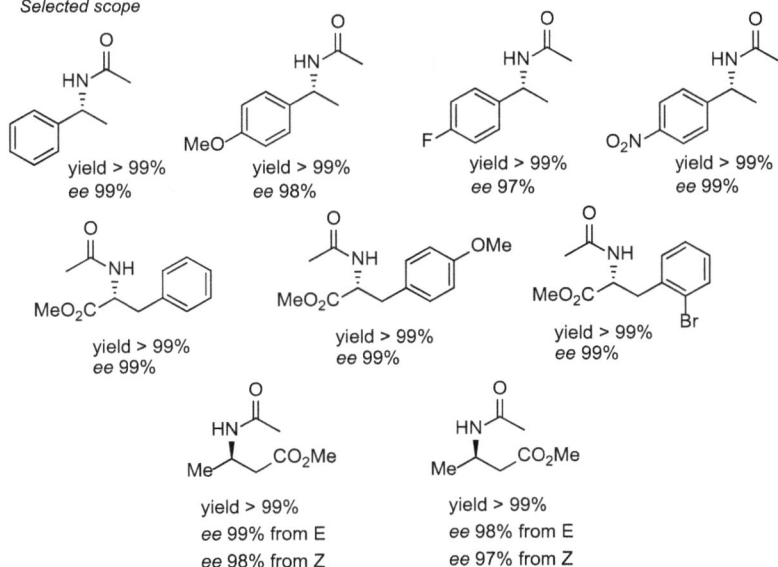

FIGURE 7.21 Continuous flow asymmetric hydrogenation of enamides and α-/β-dehydroamino acids.

The Rh-catalyst showed excellent enantioselectivity and activity for asymmetric hydrogenation of enamides **48** and **50**, and α-/β-dehydroamino acids **52** (Figure 7.21). The optimized conditions were set in a flow reactor with the time of flow (120 h^{-1}) and TON 2640 for 24 h. The chiral products were obtained with high yields and high enantiomeric ratio with *ee* > 99%. The scope of reaction with selected examples is shown in Figure 7.21. The lifetime of the catalyst was 90 h and then catalyst deactivation was detected. Up to 90 h, the TON of the enantiopure product reached 10,800 which was found to be ten times higher than homogeneous catalysts.

7.7 MIXED METAL-BASED HETEROGENEOUS CATALYSTS

The catalytic reduction of inner alkynes generally gives stereoselectivity to (*Z*)-alkenes, because of their *syn* style addition. However, recently T. Komatsu [27] reported a tandem catalytic system that comprised of intermetallic Pd_3Pb/SiO_2 (for semi-hydrogenation of alkynes) and $RhSb/SiO_2$ (for alkene isomerization from Z-alkenes to E-alkenes). The reaction was described in two steps: firstly, the half-hydrogenation of alkyne to *cis*-alkene and then its isomerization to thermodynamically more stable *trans*-alkene (Figure 7.22).

The reaction was tolerated to a variety of functionalized alkynes with aldehyde, ketone, carboxylic acid and ester groups (Table 7.5). The first step of the given hydrogenation reaction of alkyne **54** gives the usual Z-alkene **55** in the presence of a Pd-catalyst which is followed by the isomerization by Rh-catalyst. The second step, isomerization is quite difficult as it needs rotation of double bond to isomerize *cis*-alkene **55** to *trans*-alkene **56** (Figure 7.22). However, the geometric constraints of the 1-D planes of $RhSb/SiO_2$ catalyst system and steric hindrance from one alkyl group of Z-alkene limit hydrogen access toward the alkenyl carbon to one direction. This enables one-atom hydrogenation for isomerization but inhibits two-atom hydrogenation preventing overhydrogenation to alkane.

Carboxylic acids are an important class of organic compounds and these building blocks are abundantly available from biomass resources (*e.g.*, fatty acids); however, their hydrogenation has been found to be more difficult. The reduction of carboxylic acids produces alcohols that find numerous applications in organic chemistry and

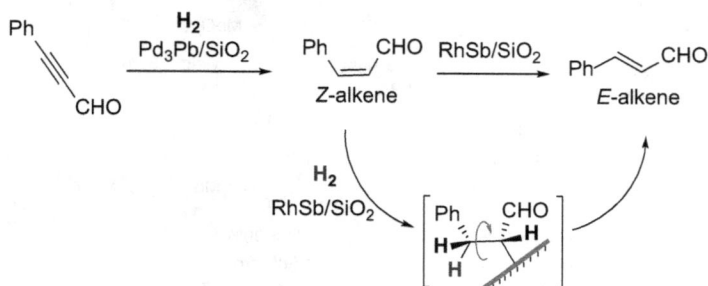

FIGURE 7.22 Isomerization of *Z*-alkene to *E*-alkene in a tandem process using Pd-Rh catalyst system.

TABLE 7.5
The Scope of the Hydrogenation of Alkynes with Pd₃Pb/SiO₂ and RhSb/SiO₂

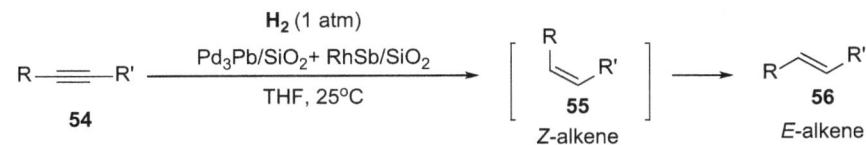

$$R\!\!=\!\!R' \xrightarrow[\text{THF, 25°C}]{\substack{H_2\,(1\text{ atm})\\ Pd_3Pb/SiO_2+\ RhSb/SiO_2}} \left[\ \underset{\substack{\textbf{55}\\ \text{Z-alkene}}}{R'}\ \right] \longrightarrow \underset{\substack{\textbf{56}\\ \text{E-alkene}}}{R'}$$

54

Substrate	Time (h)	Conversion (%)	Selectivity (E to Z) (%)	E:Z
3-Phenylpropiolaldehyde	4.3	99	95	>99:1
3-Phenylpropiolic acid	7	>99	92	>99:1
3-Phenylprop-2-yn-1-ol	12	88	84	>99:1
1,2-Diphenylethyne	2.3	>99	86	97:3
Ethyl oct-2-ynoate	0.7	99	89	97:3

chemical industries. However, carboxylic acids are hydrogenated to alcohols by the use of stoichiometric amounts of metal hydrides which imposes safety issues and waste generation. Breit and co-workers have recently reported a bimetallic system (Pd-Re) supported over high surface area graphite (HSAG) to chemoselective reduction of carboxylic acids to alcohols or alkanes [28]. Firstly, they optimized reaction conditions for chemoselective reduction of stearic acid and then they explored the scope of reaction with various saturated and unsaturated carboxylic acids to reduce to corresponding alcohols. Later, they applied optimized reaction conditions to reduce amides and esters also. It was revealed that under the given conditions of hydrogenation with Pd-Re/C catalytic system, the ease of hydrogenation of carboxylic acid followed the order: $RCO_2Et < RCONH_2 < RCOOH$ which is different than the usual reactivity order of hydrogenation: $RCONH_2 < RCOOH < RCO_2Et$. Also, regioselectivity was observed when phenylacetic acid was reduced to the corresponding alcohol where phenyl ring was not reduced, whereas, in the case of benzoic acid, both phenyl ring as well as –COOH group were reduced. Similar results were obtained for compounds with –COOH group directly attached to aryl rings. Interestingly, enantiomerically rich tetrahydrofuranyl was obtained from its respective alcohol without losing considerable optical purity.

The reaction conditions were further optimized for chemoselective outcome of the reaction with the same catalytic system to reduce fatty acids 57 to alcohols 58 or alkanes 59 (Figure 7.23). It was found that long-chained fatty acids were derivatized to alcohols or alkanes depending upon the temperature and pressure of hydrogen gas. At lower temperature and H_2-pressure, alcohol product 58 was formed and at higher temperature and H_2-pressure, the alkane product 59 was favored. The scope of the reaction is described in Figure 7.23 with the representative examples.

The author also utilized the same bimetallic catalytic system Pd-Re/C for the reduction of secondary and tertiary amides to their corresponding amines [28b].

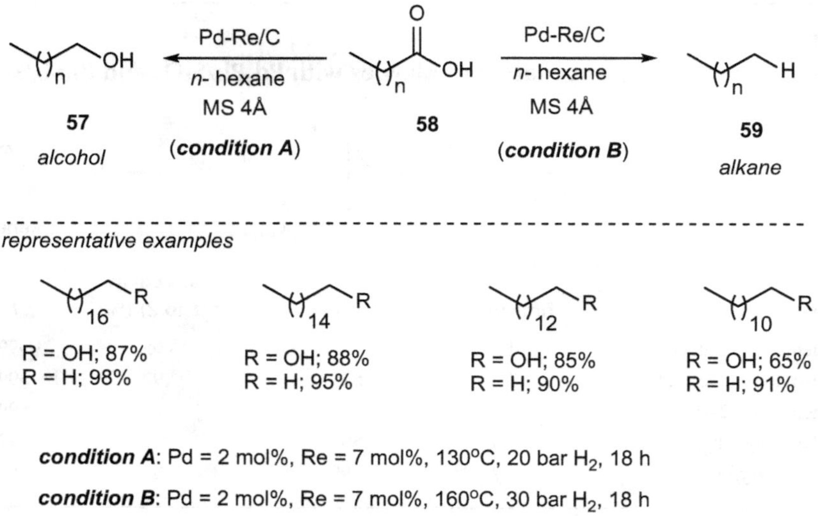

representative examples

condition A: Pd = 2 mol%, Re = 7 mol%, 130°C, 20 bar H$_2$, 18 h

condition B: Pd = 2 mol%, Re = 7 mol%, 160°C, 30 bar H$_2$, 18 h

FIGURE 7.23 Chemoselective reduction of acids using Pd-Re/C.

The amides are considered to be the least reactive and a harsh condition of temperature and pressure is required to reduce them to amines. However, with Pd-Re/C system, the reaction worked smoothly and it was able to produce a big scope of the reaction (108 examples) with a variety of amides as substrates. It is worth noting that, the reaction rate could be adjusted by changing the temperature depending on the steric hindrance of the amide substrate. This allowed even very hindered substrates to be hydrogenated in high yields. The workup of the reaction was easy, and by simple filtration, the desired amine could be purified without significant impurities and it also allowed to recycle the catalyst, which could be used in two subsequent runs with a loss of about 20% and 30% activity, respectively. The methodology was simple and environmentally friendly that provides an important tool to convert amides to amines.

Hydrotalcites (HT) belong to the class of basic clays that have anionic *layered double hydroxides* (LDH) consisting of positively charged layers resembling Brucites and compensating anions like CO_3^{2-}, HCO_3^-, OH^- along with water molecules. HT have the general formula $[M^{2+}_{1-x}M^{3+}_x(OH)_2]^{x+}A^{n-}_{x/n} \cdot mH_2O$, where ($0 < x < 1$), M^{2+} is a divalent metal (Mg, Ni, Co, Zn) and M^{3+} is trivalent metal (Al, Cr, Fe, In), x represents the fraction of the M^{3+} cation, *i.e.*, $x = M(III)/(M(III) + M(II))$, and A^{n-} denotes anions. Under the high-temperature calcination process, there is a loss of anions and water molecules, and the resulting solid solution can have varying proportions of mixed metal oxides (MMOs).

These MMOs can be obtained in different molar compositions and they act as potential heterogeneous catalysts in organic chemistry. Moreover, HT materials can also be modified by partial or full replacement of metal ions or inorganic/organic

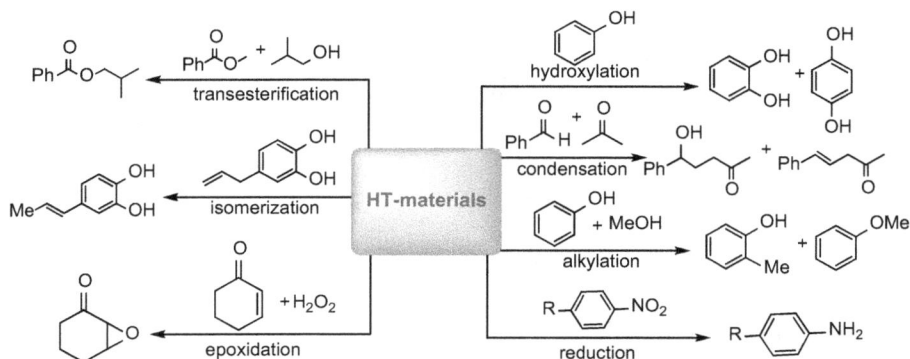

FIGURE 7.24   General applications of HT-materials in organic synthesis.

anions producing a synthetic form of HT-like materials. Apart from catalysts in organic synthesis, HT materials are also used as ion-exchangers, adsorbents, corrosion inhibitors, electrode materials and pharmaceuticals. In organic chemistry, the general applications of HT materials in various organic transformations are shown in Figure 7.24 [29].

Group of G. Prieto recently reported [30] a selective tandem isomerization-hydrosilylation of olefins by combining Rh- and Ru-based SACs, Rh_1/CeO_2 and Ru_1/CeO_2. The monoatomic catalysts individually showed different activity, Ru_1/CeO_2 SAC specifically produced internal alkenes (isomerized products) of olefin substrates (1- or 2-octene), whereas Rh_1/CeO_2 was specific for hydrosilylation in *anti*-Markovnikov manner producing a terminal organosilane product. The combined use of two SACs in a one-pot process was successfully employed for various internal alkenes (including a mixture of isomers) to convert them into terminal hydrosilylated products with good to excellent yields and regioselectivity (>95%, high terminal-to-branch molar ratio). These results have shown the significance of metal-heterogeneous SACs in tandem catalytic reactions.

G. Liu [31] recently published a report of bifunctionalized periodic mesoporous organosilica (PMO) based organopalladium/organoruthenium catalyst system utilized for a cascade process of asymmetric hydrogenation and Suzuki-coupling sequence in a one-pot reaction (Figure 7.25).

The reaction produced asymmetric alcohol with high enantiomeric excess (*ee* > 95%) and excellent yields (99% in all cases, selected scope of the reaction is shown in Figure 7.25). The Authors demonstrate the plausible sequence by the time course of reaction as the first asymmetric hydrogenation of acetophenone derivative **60** occurs that was catalyzed by Ru-catalyst and then Suzuki cross-coupling takes place between aryl-alcohol **62** and aryl-boronic acid **61** in the presence of Pd-catalyst providing the target molecules **63** (Figure 7.25). Moreover, the catalyst could be recovered easily and reused up to eight times without considerable loss in the catalyst activity and outcome of the reaction.

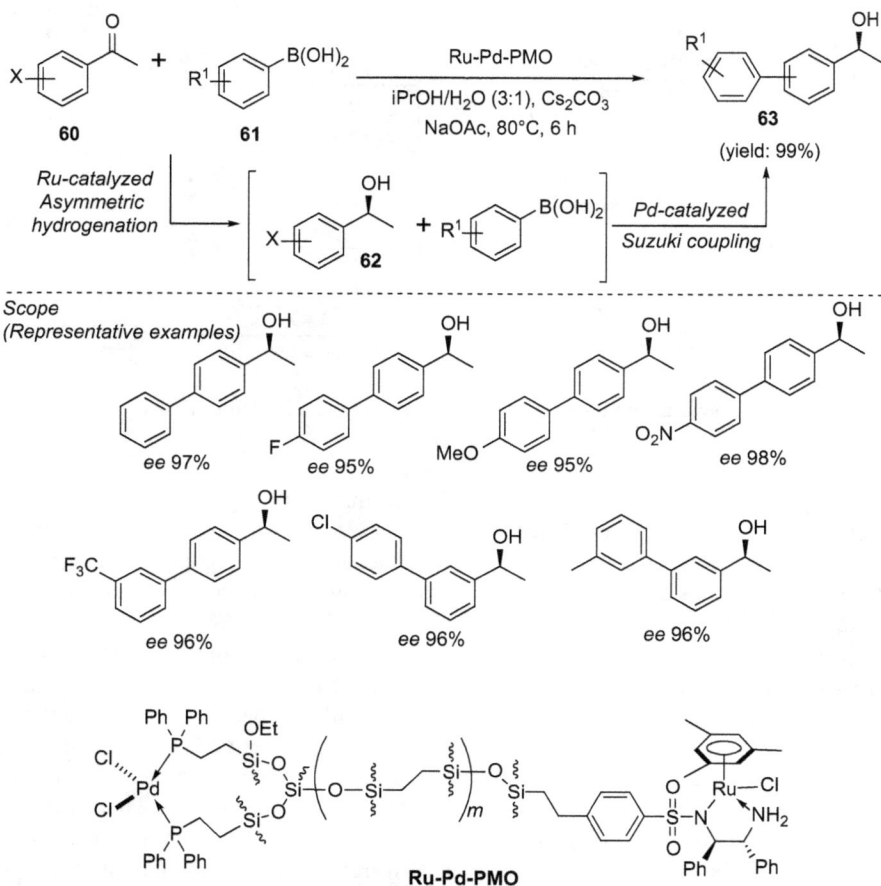

FIGURE 7.25 One-pot asymmetric hydrogenation and Suzuki-coupling reaction.

REFERENCES

1. Ma, Z. & Zaera, F. (2014). Heterogeneous Catalysis by Metals. In R.A. Scott (Ed.), *Encyclopedia of Inorganic and Bioinorganic Chemistry* (pp. 1–16). Wiley, New Jersey, United States.
2. Smith, G. V. & Notheisz, F. (1999). Introduction to Catalysis. In Smith, G. V. & Notheisz, F. (Eds.), *Heterogeneous Catalysis in Organic Chemistry* (pp. 1–26), Academic Press, Massachusetts, United States.
3. Shakeri, M., Tai, C.-W., Gçthelid, E., Oscarsson, S., & Bäckvall, J.-E. (2011). Small Pd Nanoparticles Supported in Large Pores of Mesocellular Foam: An Excellent Catalyst for Racemization of Amines. *Chemistry—A European Journal*, 17, 13269–13273. doi:10.1002/chem.201101265
4. Li, M.-B., Posevins, D., Gustafson, K. P. J., Tai, C.-W., Shchukarev, A., Qiu, Y., & Bäckvall, J.-E. (2019). Diastereoselective Cyclobutenol Synthesis: A Heterogeneous Palladium-Catalyzed Oxidative Carbocyclization-Borylation of Enallenols. *Chemistry—A European Journal*, 25, 210–215. doi:10.1002/chem.201805118

5. Li, M.-B., Inge, A. K., Posevins, D., Gustafson, K. P. J., Qiu, Y., & Bäckvall, J.-E. (2018). Chemodivergent and Diastereoselective Synthesis of γ-Lactones and γ-Lactams: A Heterogeneous Palladium-Catalyzed Oxidative Tandem Process. *Journal of the American Chemical Society*, 140 (44), 14604–14608. doi:10.1021/jacs.8b09562
6. Li, M.-B., Posevins, D., Geoffroy, A., Zhu, C., & Bäckvall, J.-E. (2020). Efficient Heterogeneous Palladium-Catalyzed Oxidative Cascade Reactions of Enallenols to Furan and Oxaborole Derivatives. *Angewandte Chemie-International Edition*, 59, 1992–1996. doi:10.1002/anie.201911462
7. Shi, W., Zeng, L., Cao, L., Huang, Y., Wang, C., & Wenbin, L. (2021). Metal-organic layers as reusable solid fluorination reagents and heterogeneous catalysts for aromatic fluorination. *Nano Research*, 14, 473–478. doi:10.1007/s12274-020-2698-8
8. Shokrolahi, A., Zali, A., & Ghani, K. (2013). Rapid Reduction of Alkenes and Alkynes over Pd Nanoparticles Supported on Sulfonated Porous Carbon. *Journal of Chemistry*, 2013, 1–7, Article ID 268649. doi:10.1155/2013/268649
9. Ibhadon, A.O., & Kansal, S.K. (2018). The Reduction of Alkynes over Pd-based Catalyst Materials-A Pathway to Chemical Synthesis. *Journal of Chemical Engineering & Process Technology*, 9(2), 1–15. doi:10.4172/2157-7048.1000376
10. Wang, Z. (2010). Lindlar Hydrogenation. In Wang, Z. (Ed.) *Comprehensive Organic Name Reactions and Reagents* (pp. 1758–1762). Wiley, New Jersey, United States.
11. Delbecq, F., Loffreda, D. & Sautet, P. (2010). Heterogeneous catalytic hydrogenation: Is double bond/surface coordination necessary?. *The Journal of Physical Chemistry Letters*, 1, 323–326. doi:10.1021/jz900159q
12. Preger, Y., Root, T. W., & Stahl, S. S. (2018). Platinum-Based Heterogeneous Catalysts for Nitrile Synthesis via Aerobic Oxidative Coupling of Alcohols and Ammonia. *ACS Omega*, 3, 6091–6096. doi:10.1021/acsomega.8b00911
13. Miura, H., Wada, K., Hosokawa, S., Sai, M., Kondo, T. & Inoue, M. (2009). A heterogeneous Ru/CeO$_2$ catalyst effective for transfer-allylation from homoallyl alcohols to aldehydes. *Chemical Communication*, 4112–4114. doi:10.1039/B901830A
14. Miura, H., Wada, K., Hosokawa, S., & Inoue, M. (2010). Recyclable Solid Ruthenium Catalysts for the Direct Arylation of Aromatic C–H Bonds. *Chemistry—A European Journal*, 16, 4186–4189. doi:10.1002/chem.200903564
15. Warratz, S., Burns, D. J., Zhu, C., Korvorapun, K., Rogge, T., Scholz, J., Jooss, C., Gelman, D., & Ackermann, L. *meta*-C–H Bromination on Purine Bases by Heterogeneous Ruthenium Catalysis. *Angewandte Chemie-International Edition*, 56, 1557–1560. doi:10.1002/anie.201609014
16. (a) Mehler, T., Behnen, W., Wilken, J., & Martens, J. (1994). Enantioselective catalytic reduction of acetophenone with borane in the presence of cyclic α-amino acids and their corresponding β-amino alcohols. *Tetrahedron Asymmetry*, 5, 185–188; (b) Hasirci, V. N. (1982). PVNO—DVB hydrogels: Synthesis and characterization. *Journal of Applied Polymer Science*, 27(1), 33–41; (c) Blickenstaff, R. T., Hanson, W. R., Reddy, S., & Witt, R. (1995). Potential radioprotective agents—IV. Schiff bases. *Bioorganic & Medicinal Chemistry*, 3, 917–922. doi:10.1016/S0968-0896(00)82087-8
17. (a) Habibi, D., Faraji, A. R., Arshadi, M., & Fierro, J. L. G. (2013). Characterization and catalytic activity of a novel Fe nano-catalyst as efficient heterogeneous catalyst for selective oxidation of ethylbenzene, cyclohexane, and benzylalcohol. *Journal of Molecular Catalysis A: Chemical*, 372, 90–99; (b) Sivaramakrishna, A., Suman, P., Goud, E. V., Janardan, S., Sravani, C., Yadav, C. S., & Clayton H. S. (2012). Recent progress in oxidation of n-alkanes by heterogeneous catalysis. *Research and Reviews in Materials Science and Chemistry*, 1, 75–103. http://www.jyotiacademicpress.net/recent_progress_in_oxidation.pdf

18. Wahlen, J., Moens, B., De Vos, D. E., Alsters, P. L., Jacobs, P. A. (2004). Titanium Silicalite 1 (TS-1) Catalyzed Oxidative Transformations of Furan Derivatives with Hydrogen Peroxide. *Advance Synthesis & Catalyst*, 346, 333–338. doi:10.1002/adsc. 200303185

19. Liua, X., Liub, J., Xiaa, Y., Yina, D., Stevena, R. K., & Mao, L. (2019). Catalytic performance of TS-1 in oxidative cleavage of 1-alkenes with H_2O_2. *Catalysis Communications*, 126, 40–43. doi:10.1016/j.catcom.2019.04.021

20. Rodenas, Y., Mariscal, R., Fierro, J. L. G., Alonso, D. M., Dumesic, J. A. & Granados, M. L. (2018). Improving the production of maleic acid from biomass: TS-1 catalysed aqueous phase oxidation of furfural in the presence of γ-valerolactone. *Green Chemistry*, 20, 2845–2856. doi:10.1039/C8GC00857D

21. Vila C., & Rueping, M. (2013). Visible-light mediated heterogeneous C–H functionalization: oxidative multi-component reactions using a recyclable titanium dioxide (TiO_2) catalyst. *Green Chemistry*, 15, 2056–2059. doi:10.1039/C3GC40587G

22. Zoller, J., Fabry, D. C., & Rueping, M. (2015). Unexpected Dual Role of Titanium Dioxide in the Visible Light Heterogeneous Catalyzed C–H Arylation of Heteroarenes. *ACS Catalysis*, 5, 3900–3904. doi:10.1021/acscatal.5b00668

23. Hanf, S., Rupflin, L. A., Gläser, R., & Schunk, S. A. (2020). Current State of the Art of the Solid Rh-Based Catalyzed Hydroformylation of Short-Chain Olefins. *Catalysts*, 10(5), 510. doi:10.3390/catal10050510

24. Amsler, J., Sarma, B. B., Agostini, G., Prieto, G., Plessow, P. N., & Studt, F. (2020). Prospects of Heterogeneous Hydroformylation with Supported Single Atom Catalysts. *Journal of the American Chemical Society*, 142, 5087–5096. doi:10.1021/jacs.9b12171

25. Lang, R., Li, T., Matsumura, D., Miao, S., Ren, Y., Cui, Y.-T., Tan, Y., Qiao, B., Li, L., Wang, A., Wang, X., & Zhang, T. (2016). Hydroformylation of Olefins by a Rhodium Single-Atom Catalyst with Activity Comparable to $RhCl(PPh_3)_3$. *Angewandte Chemie-International Edition*, 55, 16054–16058. doi: 10.1002/anie.201607885

26. Saito, Y., & Kobayashi, S. (2020). Development of Robust Heterogeneous Chiral Rhodium Catalysts Utilizing Acid–Base and Electrostatic Interactions for Efficient Continuous-Flow Asymmetric Hydrogenations. *Journal of the American Chemical Society*, 142, 39, 16546–16551. doi:10.1021/jacs.0c08109

27. Furukawa, S., & Komatsu, T. (2016). Selective Hydrogenation of Functionalized Alkynes to (E)-Alkenes, Using Ordered Alloys as Catalysts. *ACS Catalysis*, 6, 2121–2125. doi:10.1021/acscatal.5b02953

28. (a) Ullrich, J., & Breit, B. (2018). Selective Hydrogenation of Carboxylic Acids to Alcohols or Alkanes Employing a Heterogeneous Catalyst. *ACS Catalysis*, 8, 785–789; (b) Stein, M., & Breit, B. (2013). Catalytic Hydrogenation of Amides to Amines under Mild Conditions. *Angewandte Chemie-International Edition*, 52, 2231–2234. doi:10.1002/ange.201207803

29. (a) Kaneda, K., & Mizugaki, T. (2019). Design of high-performance heterogeneous catalysts using hydrotalcite for selective organic transformations. *Green Chemistry*, 21, 1361–1389; (b) Nishimura, S., Takagaki A., & Ebitani, K. (2013). Characterization, synthesis and catalysis of hydrotalcite-related materials for highly efficient materials transformations. *Green Chemistry*, 15, 2026–2042; (c) Sels, B. F., De Vos, D. E., & Jacobs, P. A. (2001). Hydrotalcite-like anionic clays in catalytic organic reactions. *Catalysis Reviews*, 43, 443–488; (d) Baskaran, T., Christopher, J., & Sakthivel, A. (2015). Progress on layered hydrotalcite (HT) materials as potential support and catalytic materials. *RSC Advances*, 5, 98853–98875. doi:10.1039/C5RA19909C

30. Sarma, B. B., Kim, J., Amsler, J., Agostini, G., Weidenthaler, C., Pfänder, N., Arenal, R., Concepción, P., Plessow, P., Studt, F., & Prieto, G. (2020). One-Pot Cooperation of Single-Atom Rh and Ru Solid Catalysts for a Selective Tandem Olefin Isomerization-Hydrosilylation Process. *Angewandte Chemie-International Edition*, 59, 5806–5815. doi:10.1002/anie.201915255

31. Zhang, D., Xu, J., Zhao, Q., Cheng, T., & Liu, G. (2014). A Site-Isolated Organoruthenium-/Organopalladium-Bifunctionalized Periodic Mesoporous Organo-silica Catalyzes Cascade Asymmetric Transfer Hydrogenation and Suzuki Cross-Coupling. *ChemCatChem*, 6, 2998–3003. doi:10.1002/cctc.201402445

Index

For Product Safety Concerns and Information please contact our EU
representative GPSR@taylorandfrancis.com
Taylor & Francis Verlag GmbH, Kaufingerstraße 24, 80331 München, Germany

www.ingramcontent.com/pod-product-compliance
Lightning Source LLC
Chambersburg PA
CBHW070713220326
41598CB00024BA/3131